Out
Of their
Minds

DENNIS SHASHA
CATHY LAZERE

OUT OF THEIR MINDS

*The Lives and Discoveries of
15 Great Computer Scientists*

COPERNICUS
AN IMPRINT OF SPRINGER-VERLAG

Published in the United States by Copernicus,
an imprint of Springer-Verlag New York, Inc.
175 Fifth Avenue
New York, NY 10010

Library of Congress Cataloging-in-Publication Data

Shasha, Dennis Elliott.
    Out of their minds : the lives and discoveries of 15 great computer
scientists / Dennis Shasha, Cathy Lazere.
      p.   cm.
    "Copernicus imprint."
    Includes bibliographical references and index.
    ISBN 0-387-98269-8 (softcover: alk. paper)
    1. Computer scientists—Biography.   2. Computer science—History.
I. Lazere, Cathy A.   II. Title.
QA76.2.A2S53   1998
004'.092'2—dc21
[B]   98-16911

Manufactured in the United States of America.
Printed on acid-free paper.

9 8 7 6 5 4 3 2 1

ISBN 0-387-98269-8   SPIN 10557986

To our families—
who saved us from going out of our minds:

The Lazeres: Monroe, Muriel, Eric, Maria, Brian, and Keith

The Shashas: Alfred, Hanina, Carol, Joe, Robert, Ellen, Karen, Cloe, Tyler, Jeff, Ariana, Nick, Jordan, and Carrie

CONTENTS

Introduction

In most sciences, the seminal thinkers lived in the remote past. To uncover what they did and why they did it, we must scavenge in the historical record, picking among scraps of information, trying to separate facts from mythology: Did an apple really fall on Newton's head? Did Archimedes jump out of his bathtub and shout "Eurcka"? Did Euclid steal the secrets of geometry from Egyptian priests while he lived in Egypt?

Computer science is different. The mathematicians who first studied computation in its current form—Alan Turing, Emil Post, and Alonzo Church—did their work in the 1930s and 1940s. Their conception of the computer is the one we still live with: a calculating engine and a memory for storing instructions as well as data. They posed the fundamental theoretical question for the field: What can be computed in finite time? To these mathematicians, a "computer" could be a human being as easily as a machine—computing machines that stored instructions did not exist when they began theorizing.

Once computers became an engineering reality in the late 1940s, the important questions in computer science took a pragmatic turn. It was not enough to know that a problem could be solved eventually. Solving it efficiently and, if possible, elegantly became the issue. This required writing instructions for a real machine, finding an efficient solution, building a better computer for bigger versions of the problem, and, sometimes, asking the computer to participate in the creative process. The four parts of this book reflect the four basic questions computer scientists have wrestled with over the last fifty years:

1. Linguists: How should I talk to the machine?
2. Algorithmists: What is a good method for solving a problem fast on my computer?
3. Architects: Can I build a better computer?
4. Sculptors of Machine Intelligence: Can I write a computer program that can find its own solutions?

This book presents the lives and work of fifteen of the greatest living computer scientists. They include eight Turing Award winners, the equivalent of the Nobel Prize for computer science, and all are innovators of first rank. In fact, modern computing would be unrecognizable without their contributions. In the following pages, they explain how they became interested in science, how they were influenced by other scientists and their environment, how they arrived at their basic discoveries, and what their vision of the future is.

We try to explain the ideas and their importance without scientific jargon, so you need not have any special background, other than curiosity about how computing has evolved and how this special breed of scientists thinks.

Imagine visiting Isaac Newton in 1690. You might ask for his views about inertial forces and he might tell you his memories of farm life in Woolsthorpe. It is the privilege of talking with the living Isaac Newtons of computer science that inspired us to write *Out of Their Minds*.

A Note on Style and Presentation

We have based this book on in-person interviews, because we believe you will get a clearer picture of the scientists' personalities and methods from reading their own words in conversation than from reading our interpretations. We have tried to set the historical background and explain the underlying science and mathematics, but let the scientists tell you directly how they made their discoveries and what they were thinking about when they made them.

We have relegated detailed technical points to boxes, considering them mostly of interest to the cognoscenti. You may want to come back to them on a second reading. If you're new to the field, we have

also included a glossary of basic terms. To help you orient these profiles historically, we have included a timeline on the endpapers of
major dates in the field and in these scientists' lives. The last two
chapters, the Epilogue and Postscript, explore two questions: Is there
a family resemblance in the lives and thoughts of great computer scientists; and twenty-five years from now, which discoveries will a
book like this discuss?

$\mathcal{L}$INGUISTS
HOW TO TALK TO MACHINES

Imagine that you are the chief engineer of a computer construction project shortly after World War II. You have collected a bunch of wires and switches and your mission is to construct a machine that predicts the patterns of planes flying through the air. Because of the technology of the time, you must build this machine out of unreliable parts and engineer it to be easily fixed. It will consume enough electricity to run a small factory. You endow it with a memory big enough to hold a few thousand characters, an adder, a multiplier, and a set of instructions for the calculations.

The instructions are the least of your worries. They only need to be expressive enough to get the job done.

For efficiency's sake, you specify instructions that reflect much of the detail of the engineering design. Your machine has memory of different speeds and expense. You therefore come up with a group of instructions that moves data between slow memory and fast memory and another group of instructions to add, subtract, multiply, divide, and otherwise manipulate data once it's in fast memory. This grouping makes it more difficult to write a program to predict flight patterns than a grouping that treats memory as an undifferentiated resource, but this doesn't worry you. Mathematician-programmers come a dime a dozen. The machine is delicate, expensive, and unique.

Five years later, the situation has changed completely. To your surprise, you find that the surging demand for programs and the declining cost of hardware make programming more expensive than buying and maintaining the hardware. (This continues to be the case. As of the mid-1990s, programming costs are roughly fifty times hardware costs in most corporate software development settings.) So you set yourself two new goals: reduce the time required to produce the first version of a program and make the program easier to improve when customers request changes.

You decide to change the language that programmers use to communicate to the machine. Instead of reflecting the memory hierarchy and arithmetic operations of the machine, the language should reflect the structure of the problem you want to solve. You know that every profession invents a language for its purposes. For example, film directors use terms like *action* as shorthand for "Cameramen, start your cameras. Actors play the current scene. Agent get off the set." Since computers serve mostly scientific and data processing applications in the mid-1950s, you focus on creating a mathematical language: arithmetic formulas, sums, and vectors. To get some ideas, you visit various researchers, and see projects that allow programmers to write arithmetic expressions like $(X + Y)/Z$ and, voilà, the computer performs the appropriate calculation.

By 1958, a clear winner among scientific programming languages emerges: Fortran (standing for *for*mula *trans*lation) designed by *John*

Backus and a few of his colleagues at IBM. The language remains the most widely used programming language of physical science.

Meanwhile, in 1956, a small group of researchers launches a new discipline within computer science: artificial intelligence. A goal for this discipline is to create a computer with human reasoning capabilities. Since much of human reasoning entails manipulating abstractions—-the relative positions of a set of building blocks, the grammar of a sentence, the pros and cons of different laundry soaps—Fortran's arithmetic bias makes it inappropriate for this purpose. Allen Newell and Herbert Simon of Carnegie Mellon suggest representing abstractions as lists of symbols and modeling reasoning as "symbol manipulation."

Imagine a playroom with a red block, a blue block, and a yellow block. A 5-year-old can tell you that if the red block is above the blue block and the yellow block is above the red block, then the yellow block is above the blue block. Representing the first two facts as lists, you might write: (above redblock blueblock) and (above yellowblock blueblock). You might then include a symbol-manipulation rule saying that (above x y) and (above z x) implies (above z y).

The young *John McCarthy* was impressed with this argument, but considered the Newell-Simon design too complex. He simplified it, added two powerful features known as *recursion* and *eval,* and created a language called Lisp (for *list* processing). Besides being the language of choice for artificial intelligence, Lisp, together with Fortran, has provided the inspiration for computer language designs ever since its introduction in 1958.

In the summer of that year, an international group of computer scientists met in Zurich to design a successor to Fortran that would incorporate elements of Lisp. Their collaborative effort led to Algol 60, the parent of languages such as Pascal, C, and Ada and the ancestor of languages ranging from Postscript to Excel.

By the early 1970s, however, a small group of researchers had come to believe that the hundreds of languages derived from Fortran,

Algol, and Lisp had a common failing: it was nearly impossible to understand a program once it was written. Practitioners found that changing the behavior of a program often entailed rewriting it.

Drawing upon his background in biology, *Alan C. Kay* concluded that programs should be built like living creatures: out of autonomous "cells" that cooperate by sending messages to one another. He reasoned that building programs out of autonomous units would permit those units to be used in new contexts. For *cell* he used the word *object* and called the approach *object-orientation*. The object's behavior is defined by the messages it responds to.

An application concerned with images, for example, would describe their common behavior through messages or commands such as "rotate 30 degrees" or "print yourself." A program controlling moon-walking robots, by contrast, might specify that robots respond to messages like "collect rock sample" or "locate boulders over 1 meter high." Kay and his team at Xerox designed the first object-oriented language, called Smalltalk, in the mid-1970s. Young teenagers were among its first users a few years later. Since then, Smalltalk has profoundly influenced such decidedly grown-up languages as C++ and has a professional following of its own.

In the past forty years, thousands of languages have been designed; around 60 remain in widespread use, including special-purpose ones for word processing, spreadsheets, graphics, and animation. A good language can help solve a computer problem in the same way a tool helps solve a carpenter's problem. And programmers become as attached to their favorite languages as carpenters do to their tools.

The linguist Benjamin Whorf once wrote "Language shapes the way we think, and determines what we can think about." Human languages inspired those words, but computer languages—especially those inspired by the work of Backus, McCarthy, and Kay—confirm his assertion every millisecond.

John Backus

A RESTLESS INVENTOR

We didn't know what we wanted and how to do it. It just sort of grew. The first struggle was over what the language would look like. Then how to parse expressions—it was a big problem and what we did looks astonishingly clumsy now. . . .

—JOHN BACKUS ON THE INVENTION OF FORTRAN

ecessity, says the adage, is the mother of invention. Yet some inventors are motivated less by necessity than by sheer irritation at imprecision or inefficiency. John Backus is such an inventor. He played an inspirational role in three great creations: Fortran, the first high-level programming language; Backus-Naur form, which provides a way to describe grammatical rules for high-level languages; and a functional programming language called FP. Today, each of these inventions drives research and commercial agendas throughout the world. Yet for Backus himself, the inventions arose because of his impatience with the conceptual tools he found available.

A distaste for inefficiency seems to run in the family. Before World War I, Backus's father had risen from a modest background to the post of chief chemist for the Atlas Powder Company, a manufacturer of nitroglycerine used in explosives. His promotion came for good reason.

Their plants kept blowing up or having very poor yields and they couldn't figure out why. The yield was very temperature sensitive. My father discovered that the very expensive German thermometers they had were incorrect. So, he went to Germany and studied thermometer making and got some good thermometers and their plants stopped blowing up so much.

During World War I, Backus senior served as a munitions officer. A promised postwar job at DuPont never materialized, so he became a stockbroker instead. By the time John Backus was born in Philadelphia in 1924, his father had grown rich in the postwar boom. Backus spent his early years in Wilmington, Delaware, and attended the prestigious Hill School in Pottstown, Pennsylvania.

> I flunked out every year. I never studied. I hated studying. I was just goofing around. It had the delightful consequence that every year I went to summer school in New Hampshire where I spent the summer sailing and having a nice time.

After a delayed graduation from the Hill School in 1942, Backus attended the University of Virginia, where his father wanted him to major in chemistry. Backus liked the theory, but hated the labs. He spent most of his time at parties, waiting to be drafted. By the end of his second semester, Backus was attending only one class a week—an untaxing music appreciation class. Finally, the school authorities caught up with him and his career at the University of Virginia ended. He joined the Army in 1943.

Backus became a corporal in charge of an antiaircraft crew at Fort Stewart, Georgia, but his performance on an aptitude test convinced the Army to send him to a preengineering program at the University of Pittsburgh. A later medical aptitude test may have saved his life.

> My friends were shipped off to the Battle of the Bulge and I went to Haverford College to study premed.

As part of the premed program, Backus worked at an Atlantic City hospital in a neurosurgery ward that treated head wounds. In a bizarre coincidence, Backus was diagnosed with a bone tumor and a plate was installed in his head. Soon after, he attended medical school at Flower and Fifth Avenue Hospital (now New York Medical College), but that lasted only nine months.

> I hated it. They don't like thinking in medical school. They memorize—that's all they want you to do. You must not think.

Backus also found out that the metal plate in his head did not fit properly, and went to a nearby Staten Island hospital that specialized in plates to have a replacement made. Not satisfied with the proposed design, he got to know the technicians and designed his own. After that, Backus quit the medical field. He took a small apartment in New York City for 18 dollars a month.

> I really didn't know what the hell I wanted to do with my life. I decided that what I wanted was a good hi fi set because I liked music. In those days, they didn't really exist so I went to a radio technicians' school. I had a very nice teacher—the first good teacher I ever had—and he asked me to cooperate with him and compute the characteristics of some circuits for a magazine.
>
> I remember doing relatively simple calculations to get a few points on a curve for an amplifier. It was laborious and tedious and horrible, but it got me interested in math. The fact that it had an application—that interested me.

Backus enrolled at Columbia University's School of General Studies to take some math courses. He disliked calculus but enjoyed algebra. By the spring of 1949, the 25-year-old Backus was a few months from graduating with a B.S. in mathematics, still without any idea what to do with his life.

One day that spring, Backus visited the IBM Computer Center on Madison Avenue. He was taken on a tour of the Selective Sequence Electronic Calculator (SSEC), one of IBM's early electronic (vacuum tube) machines.

The SSEC occupied a large room, and the huge machine bulged with tubes and wires. While on the tour, Backus mentioned to the guide that he was looking for a job. She told Backus to talk to the director.

> I said no, I couldn't. I looked sloppy and disheveled. But she insisted and so I did. I took a test and did OK.

Backus was hired to work on the SSEC. The machine was actually not a computer in the modern sense. It had no memory to store software, so programs had to be fed in on punched paper tape. With its thousands of electromechanical parts, the SSEC was not too reliable either.

> It was fun to work on that machine. You had the whole thing to yourself. You had to be there because the thing would stop running every three minutes and make errors. You had to figure out how to restart it.

The state of programming was just as primitive.

> You just read the manual and got the list of instructions and that was all you knew about programming. Everybody had to figure out how to accomplish something and there were of course a zillion different ways of doing it and people would do it in a zillion different ways.

Backus worked on the SSEC for three years. His first major project was the calculation of a lunar ephemeris—the position of the moon at any given moment. In those days, IBM was able to afford the luxury of a pure science department, whose projects included a joint effort, with Columbia University, to find ways to use punched card and paper tape machines for scientific research. The firm's bread and butter was the production of punched card machines for businesses and government. The extremely visible SSEC machine was simply good public relations for a technology that even IBM would finally abandon. Whatever the company's motivations, Backus learned a great deal from his work on the SSEC. He also made his first contribution to scientific computing, Speedcoding, born of the pain of computing with large numbers on a small machine.

Speedcoding: Big Numbers in Small Words

A computer "word" is the size of the numbers it can add, subtract, and multiply. These words have a fixed length, usually ranging from 8 to 32 decimal digits, depending on the hardware. This size limit makes it necessary to perform special tricks to represent values ranging from angstroms (a ten millionth of a millimeter) to light-years.

One of the pioneers in scientific calculation was the renowned mathematician John von Neumann (1903–1957). As a child in Hungary, von Neumann was a math prodigy, completing his first mathematics paper while still in high school. He also had a remarkable memory. For fun, he once memorized the entire *Cambridge History of the Ancient World*. His research achievements included fundamental results in classical mathematics and quantum mechanics and the invention of game theory. Von Neumann emigrated to the United States in 1933, to Princeton's Institute for Advanced Study, and took a leading role in the Manhattan Project during World War II.

Von Neumann recognized that designing an atomic bomb would require calculations involving very large and very small numbers, and was intrigued when he found out about an electronic calculator project at the Moore School of Engineering at the University of Pennsylvania. He joined the project in 1944, first as an observer and later as a participant, and advocated the idea of building a machine that stored instructions as well as data—the approach used in virtually all computational devices today.[1]

After the war, von Neumann designed his own computer (universally known as the Johniac) and some of the first programs to answer questions in nuclear physics that could not be solved by hand. Von Neumann's interest in scientific calculation led him to suggest using *scaling factors* for storing and manipulating very large and very small numbers in a computer.

The idea is simple enough. Suppose that a computer word can store only three digits. To represent the number 517, you write 517 into that computer word and specify the scale as 0. To represent 51.7, you write 517 into that word and specify the scale to be −1; for 5170, you write 517 into the word and specify a scale of 1; for 5,170,000, you specify a scale of 4; and so on. A positive scaling factor is the number of 0s following the number, and a negative scaling factor is the number of digits to the right of the decimal point.

---

[1] Advocating an idea is not the same as inventing it. Von Neumann gave much of the credit to Alan Turing, whose theoretical "machine" also stored instructions. The credit for constructing an electronic memory goes to John Eckert of the Moore School, who used a vibrating tube of mercury to store data.

In mapping out a computation, programmers wouldn't know the exact value of the result of every calculation, but they should know (or so von Neumann believed) its scale. Yes, the idea is simple enough—if you happen to be a great mathematician. Backus, however, had a programmer's perspective.

> You had to know so much about the problem—it had all these scale factors—you had to keep the numbers from overflowing or setting big round off errors. So programming was very complicated because of the nature of the machine.

Working with Harlan Herrick at IBM, Backus created a program called Speedcoding to support computation with floating point numbers. *Floating point numbers* carry their scaling factor around with them, thus relieving the programmer from the drudgery of assigning one. Backus's experience with Speedcoding set the stage for a much greater challenge.

> Everybody was seeing how expensive programming was. It cost millions to rent machines and yet the cost of programming was as big or bigger. Programming was expensive because you had to hire many programmers to write in "assembly" or second generation languages, which were only one step removed from the binary or machine code of 0s and 1s. Assembly language was time-consuming; it was an arduous task to write commands that were specific to a particular machine and then to have to debug the multitude of errors that resulted. As a consequence, the ultimate goal of the program was often lost in the shuffle.

Fortran: The First High-Level Computer Language

In December of 1953, Backus wrote a memo to his boss at IBM, Cuthbert Hurd, suggesting the design of a programming language for the IBM 704 (which already had a floating point capability). That project became known as Formula Translation—Fortran. Its goals were straightforward.

> It would just mean getting programming done a lot faster. I had no idea that it would be used on other machines. There were hardly any other machines.

But first Backus had to overcome the persuasive arguments of von Neumann, who was then a consultant to IBM.

> He didn't see programming as a big problem. I think one of his major objections was you wouldn't know what you were getting with floating point calculations. You at least knew where trouble was with fixed point if there was trouble. But he wasn't sensitive to the issue of the cost of programming. He really felt that Fortran was a wasted effort.

Hurd approved the project anyway and von Neumann didn't fight it any further. Backus hired an eclectic team of experienced programmers and young mathematicians straight out of school. By the fall of 1954, his programming research group had a clear vision: to create a language that would make programming easier for the IBM 704.

As they saw it, designing the language was the easy part. Translating it to the language that the machine understood directly was the real challenge. The program that does this translation is called a *compiler*. Backus and his team lacked the algorithms that nowadays permit an undergraduate to design and implement a new language compiler within a semester. Specifically, they had no good way to design the heart of the compiler, the part known as the *parser*.

Parsing is the computer equivalent of diagramming sentences, an activity most of us happily forget once we leave school. To diagram a sentence, you identify the relationship between different parts of speech: determiner, noun, verb, adjective, adverb, and draw them in the form of a tree. (See Figure 1.)

In the case of a language compiler, the parser draws the tree and then translates a *high-level language* (one that is reasonably easy for people to understand) into machine language. *Machine language* consists of a sequence of instructions that are imprinted directly into the circuitry of a computer. (Different computer models often support different instructions—that's why a Mac program may not run on an IBM-compatible computer.)

The efficiency of a machine language program depends on how efficiently the designer of the program has used the very fast, short-term memory. Most modern machines have at least 16 registers, although some have thousands. This is a tiny number compared to the

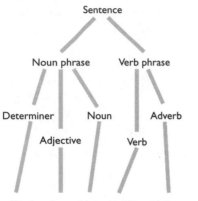

Figure 1
Parsing a sentence in English.

The Jamaican athlete skied beautifully.

millions of bytes of information in the slower RAM (random access memory). But most computational action happens in the registers.

For example, an arithmetic expression such as (A + B)/C would be translated to the following sequence:

```
copy A to register 1
copy B to register 2
copy C to register 3
add register 1 and register 2 and put result in register 1
divide register 1 by register 3 and put the result in
register 1
```

In that example, the arithmetic operations require the data to be in registers. (In some machines the data need not be in registers, but then the operation is much slower.) If the next arithmetic operation needs variables D and E, then the language translator must decide whether to put D and E in registers 4 and 5 or to reuse 1, 2, or 3. This decision in turn depends on when A, B, and C will next be needed and many other factors. It's a difficult and still-unsolved problem. Backus and his team were among the first to try to reach some reasonable solution.

But the biggest challenge had to do with a new feature of Fortran: the DO loop. These statements are used by programmers to perform repeated computations easily. For example, the sequence of instructions

```
DO 13 J = 1, 100
C[J] = A[J] + B[J]
13 CONTINUE
```

adds the first element of A to the first element of B to obtain the first element of C, then adds the second element of A to the second element of B to obtain the second element of C, and so on.

THE ARISTOCRATIC INVENTOR OF LOOPS

Unknown to Backus or any other practicing computer scientists of the 1950s, loops had been invented more than 100 years earlier. In 1833, daughter of Lord Byron, Augusta Ada, met Charles Babbage, who was designing a mechanical computational machine called the Analytical Engine. Ada, a natural mathematician from the age of 8, was one of the few to understand Babbage's vision, and they began a lifelong but properly Victorian collaboration that would make her the world's first programmer. While designing programs for the Analytical Engine, she found the need for loops and even subroutines.

She wrote up a description of the Analytical Engine, but refused to publish it under her own name, because, she argued, women were not supposed to write scientific papers. She finally agreed, under the prodding of Babbage and her husband the Earl of Lovelace, to write under the initials A. A. L. Ada's life ended tragically. She became a gambler, an alcoholic, and a cocaine addict and died of cancer at the age of 36, a fitting Byronic ending.

Compiling a DO statement efficiently requires using special registers known as *index registers*. On the IBM 704, for which Fortran was originally designed, there were only three index registers, so they were a precious resource.

Harlan [Herrick] invented this DO statement and so we knew we would have problems there—how to do it. We had three index registers and all these subscripts [J in the example above, though a complex program can have twenty or more].

It was very very difficult to figure out which index registers to use given the information in the program. You would get terrible code if you tried to do it in a general way. We knew that we would have to analyze code for its frequency and all kinds of stuff.

Backus's concern for efficiency was shared by all members of the team. Fortran's designers knew that programmers would not use their new high-level language if the machine language it generated was less efficient than what a good programmer could do by hand. For this reason, roughly half their work became directed toward generating efficient machine code. As a result, Fortran has always been known for its good performance.

The IBM 704 had only about eighty users, including General Electric, United Aircraft, Lockheed, and others in the airplane industry. By a fluke, Westinghouse became Fortran's first commercial user in April 1957, when Backus's group sent them a punched card deck containing the language's compiler.

> They just figured this was the Fortran deck and they got it to run without any instructions. They were doing hydrodynamics—calculating stresses of wing structure and stuff like that to design airplanes. Before that, they would have used desk calculators and wind tunnels.

The first application of Fortran ran without problems, but Westinghouse scientists and others soon found plenty of errors in the Fortran compiler. Backus's group fixed them over the following six months.

Although Backus had led his elite team for four years in a complicated endeavor that required considerable stamina, he is consistently modest about his participation in the project.

> It seems very unfair to me that I get so much credit for these guys [Robert Nelson, Harlan Herrick, Lois Haibt, Roy Nutt, Irving Ziller, Sheldon Best, David Sayre, Richard Goldberg, Peter Sheridan] who invented a tremendous amount of stuff. It didn't take much to manage the team. Everybody was having such a good time that my main function was to break up chess games that began at lunch and would go on to 2 p.m.

Nearly forty years after its introduction, Fortran is still the language of choice for scientific computation. A testimony to its importance is the fact that it is still evolving. For example, in 1992, an international standards committee added a new adjective to the language that allows programmers to tell the compiler that many computers may, if they are available, share the work of a single DO loop. But that jumps way ahead of our story.

Backus stopped working on Fortran in the late 1950s. His contributions to programming languages had just begun, however.

In May 1958, an international committee of prominent computer scientists from business and academia convened in Zurich. Their goal was to improve Fortran and create a single, standardized computer language. Their creation, the International Algebraic Language, later became known as Algol.

Algol had two significant advantages over Fortran. First, the new language introduced the notion of local variables. Every program names various data elements. In one of the Fortran examples above, we named the elements A, B, and C. In general, there are so many data elements in a given program that a programmer runs the risk of using the same name twice. That is, a programmer may give an already existing name to a new element in a program, resulting in some very unpleasant errors. Anyone with a common name can intuitively understand this kind of problem: bills go to the wrong address, credit is refused for the wrong reason, and late night phone calls are for the wrong person. In the programming field, names that keep their meaning over an entire program are called *global* and cases of mistaken identity are known as *name collisions.*

One way to eliminate name collisions is to localize the context of names. For example, a person speaking of "the Duke" in the city of York would normally mean the Duke of York, whereas a person speaking of "the Duke" at the Newport Jazz Festival would probably mean Duke Ellington. Similarly, a local variable is a computer memory location whose name has meaning only in a certain limited context. Outside that context, the same name may be used to designate a different memory location.

While Fortran allowed only global naming, Algol was designed with local names. Besides being convenient, local naming permits a form of programming that John McCarthy had recently introduced to computer science known as *recursion.*

A recursive computer function is defined partly in terms of itself. Let's look at an everyday example of a recursive definition. Suppose a woman defines her maternal ancestors as follows: my mother is my maternal ancestor and any maternal ancestor of my mother is my maternal ancestor. This definition may at first appear to be circular, but let's take a closer look.

Suppose Anne is the mother of Barbara who is the mother of Carol who is the mother of Donna. (We could keep on going with Eunice, Florence and so on.) According to our definition, Barbara is Carol's maternal ancestor, since she is Carol's mother. Anne is also Carol's maternal ancestor, since she is a maternal ancestor of Barbara (she is Barbara's mother). Now, that we have established that Anne is a maternal ancestor of Carol, we see that Anne must be a maternal ancestor of Donna as well.

Recursion permits a programmer to break a problem into smaller versions of itself and then glue together the solutions to each version. For example, to put a huge stack of papers in order, you might put half of the stack in order, then put the rest in order, then merge the two halves.

In a strictly theoretical sense, recursion and local naming didn't make the new language more powerful than Fortran, but they encouraged a new way of thinking that led to such later algorithmic techniques as depth-first search, which is discussed in the chapter on Tarjan.

Backus liked the ideas embodied by Algol, but felt frustrated by the difficulty of expressing them clearly.

> They would just describe stuff in English. Here's this statement— and here's an example. You were hassled in these Algol committees [with unproductive debates over terminology] enough to realize that something needed to be done. You needed to learn how to be precise.

To address this problem, Backus applied a formalism called *context-free languages* that had just been invented by linguist Noam Chomsky. (See box, "Context-Free Grammars and Backus-Naur Form" on page 18.) Chomsky's work in turn had its roots in Emil Post's theoretical work on rewriting grammars.

How Backus came to this synthesis promises to keep historians busy for some time.

> There's a strange confusion here. I swore that the idea for studying syntax came from Emil Post because I had taken a course with Martin Davis at the Lamb Estate [an IBM think tank on the Hudson]. . . . So I thought if you want to describe something, just do what Post did. Martin Davis tells me he did not teach the course until long afterward [1960–61 according to Davis's records]. So I don't know how to account for it. I didn't know anything about Chomsky. I was a very ignorant person.[2]

Backus's invention eventually became famous as Backus-Naur form due to a chain of fortuitous events that started with Backus trying to explain his ideas about precise grammars in a paper for the UNESCO meeting on Algol in Paris in June 1959.

> Of course, I had it done too late to be included in the proceedings. So I hand-carried this pile to the meeting. So it didn't get very good distribution. But Peter Naur read it and that made all the difference.

Naur was a Danish mathematician who improved Backus's notation and used it to describe all of Algol. When the programming language community started to experiment with Algol, Naur's manual proved to be the best reference available for describing the language syntax.

Liberating Programming from His Own Invention

Backus had invented one of the world's first and most popular programming languages and had developed a notation that would permit the definition of over a thousand more. Many people, even many great scientists, might have coasted after such achievements. Not Backus. He wasn't sure he liked what he had done.

---

[2]Martin Davis speculates that Richard Goldberg, a Harvard-trained logician and part of the Fortran team, may have discussed Post's or Chomsky's work with Backus.

CONTEXT-FREE GRAMMARS AND BACKUS-NAUR FORM
To gain an appreciation for Backus-Naur form, consider the following description of some English phrases. The terms *noun phrase* and *verb phrase* are borrowed from modern linguistics.

sentence → noun phrase verb phrase

noun phrase → article adjective noun | article noun

verb phrase → verb noun phrase

adjective → red | blue | yellow | big | small | smart

noun → house | girl | boy

verb → likes | hits

determiner → the | a

The vertical bar symbol | represents an alternative. For example, a determiner can be either "the" or "a." This grammar tells us to say that the sentence "the girl likes the boy" is a good sentence, since "the girl" constitutes a noun phrase, and "likes the boy" is a verb phrase. It also tells us that "a smart house likes the boy" is grammatically acceptable, even though this sentence has unusual semantics.

In programming languages, Backus-Naur form (BNF) also has this property. If you follow its rules, you will get a syntactically correct program, one that the compiler can translate successfully to machine language, preserving the meaning in the high-level language program you started with. Of course, your original program may still produce nonsense. Compilers guarantee a correct translation only, not that the original program is correct.

Once you've written a Fortran program, you can't tell what's going on really. It takes these two numbers and multiplies them and stores them here and does some other junk and then makes this test. Trying to figure out what is actually being calculated is not easy. Trying to do that calculation in a different way [is very difficult] because you basically don't understand what the program is doing.

Backus's goal was to enable programmers to state *what* they wanted to be done without getting involved with the *how*. Backus introduced this concept to the general computer science community in his 1977 Turing Award lecture entitled "Can Programming Be Liberated from the von Neumann Style?"

The reference to von Neumann had nothing to do with his earlier disagreement with the great mathematician concerning the development of Fortran. Instead, Backus was referring to von Neumann's characterization of computers as processors connected to memories where the memory contains both programs and data. Implicit in this characterization, according to Backus, is a basic cyclic process: fetching data from memory, performing some operation on it, and then returning the result to memory. Programming languages follow this paradigm at the expense of clarity, Backus believed. Building on John McCarthy's work on Lisp (mainly used in artificial intelligence) and Kenneth Iverson's language APL, Backus proposed a language called FP. A main goal of the project was to construct programs out of mathematical functions.

The main difference between *normal* (von Neumann) languages and *functional* ones like FP is that normal ones explicitly modify the computer memory whereas functional languages rely on *function composition*. Backus introduced FP in his Turing Award lecture with the example of an inner product between two arrays—an operation that is often done in physics. A standard von Neumann-style language would write this operation more or less in this form:

```
c := 0
for i := 1 step 1 until n do
    c := c + a[i] * b[i]
```

Backus criticizes this formulation on several counts, but especially the following two. First, c is repeatedly updated, so that the only way to understand the program is to understand that each update serves to add one more product to a running total. Thus, one must be able to mentally execute the code in order to understand it. Second, the formulation names its vector arguments (a and b) and their length (n). To work for general vectors, these must be passed as "reference" parameters—a tricky issue in most languages. In FP, by contrast, the inner product operation would be defined as follows:

Def Innerproduct = (Insert +)(ApplyToAll *)(Transpose)

This must be read right to left. Transpose pairs up the appropriate elements of the two vectors. In the above example, a[1] should be paired up with b[1], a[2] with b[2], and so on. "(ApplyToAll *)" replaces each pair by its product. "(Insert +)" sums up those products.

This has three principal advantages, according to Backus. First, there is no hidden state (the variable c in the program above). Second, it works for any two vectors of the same size because it does not name its arguments (thus eliminating the parameter-passing issue). Third, there is no notion of repeated calculation; each step, or *function* (Transpose, ApplyToAll, and Insert), is applied exactly once to the result of the previous step in a process known as *composition*.

Ironically, functional languages, including FP, have not caught on for one of the reasons that Fortran did: functional language programs are difficult to compile to an efficient form. As processors and memories become faster, pundits have predicted that the concern with program efficiency may diminish, but those predictions have not yet come to pass.

Also, FP is not well suited to many mundane programming tasks, particularly those that consist primarily of fetching data, updating it, and returning it to some database. Tasks such as updating an account balance are naturally suited to the von Neumann paradigm.

> Doing a functional language and combining it with the ability to do real things is very hard. And everybody has had a problem with it.

If Backus has not solved the problem, he has posed it beautifully. Other computer scientists will carry the torch from now on, however. Since his retirement in 1991, Backus has withdrawn entirely from the world of computer science, indeed from science altogether. He practices meditation and reads the introspective writings of Krishnamurti and Eva Pierrakos.

> Most scientists are scientists because they are afraid of life. It's wonderful to be creative in science because you can do it without clashing with people and suffering the pain of relationships, and making your way in the world. It's a wonderful out—it's sort of this aseptic world where you can use the very exciting faculties you have and not encounter any pain. The pain in solving a problem is small potatoes compared with the pain you encounter in living.
>
> Introspection is not a scientific activity: it's not repeatable, there are no good theories about how to do it, what you expect to find. It's strange that by looking into yourself you really get an appreciation of the mystery of the universe. You don't by trying to find the laws of physics.

John McCarthy

THE UNCOMMON LOGICIAN
OF COMMON SENSE

If you want the computer to have general intelligence, the outer structure has to be commonsense knowledge and reasoning.

—JOHN MCCARTHY

hen a 5-year-old plays with a plastic toy car, she pushes it back and forth and beeps the horn. She knows that she shouldn't roll it on the dining room table or throw it at her little brother's head. Before she leaves for school, she puts her car in a place that is too high for her brother to reach. When she returns from school, she expects to find her car just where she left it.

The reasoning that guides her actions and her expectations is so simple that any child her age could understand it. Yet most computers can't. Part of the computer's problem has to do with its lack of knowledge about day-to-day social conventions that the 5-year-old has learned from her parents, such as don't scratch the furniture and don't injure little brothers. Another part of the problem has to do with a computer's inability to reason as we do daily, using a common-sense system of educated guesses that's foreign to conventional logic and therefore to the thinking of the average computer programmer.

Conventional logic uses a form of reasoning known as *deduction*. Deduction permits us to conclude from statements such as "All unemployed actors are waiters" and "Sebastian is an unemployed actor" the new statement that "Sebastian is a waiter." The main virtue of deduction is that it is *sound*—if the premises hold, then so will the conclusions. It is also *monotonic* (a mathematical term whose root meaning is "unvarying"). If you learn new facts that do not contradict the premises, then the conclusions will still hold.

But while most of us learn deduction in school, we rarely use it in practice. The five-year-old believes her toy car will remain in its place, because she put it outside her brother's reach, but she might not feel so sure if she had seen him climb on a chair as she left for school that day. The commonsense reasoning of a 5-year-old relies on educated guesses that may have to be revised nonmonotonically as new facts surface. Not that 5-year-olds are special in this regard either.

Even Sherlock Holmes, the putative master of deduction, didn't really use deduction very often. In one of his most ingenious insights, during the adventure about an injured racehorse, *Silver Blaze*, Holmes concluded that the watchdog never barked because he knew the culprit. This is clever and plausible, and turns out to be true in the story, but it is not deduction—the dog could have been drugged, muzzled, or off hunting for rabbits.

Programmers know how to get a computer to perform deduction, because the mathematics involved is well understood. But if you want a computer to perform the conjectural—but usually correct—commonsense reasoning upon which human survival depends, you must invent a whole new kind of mathematical logic. That's one of the goals John McCarthy set for himself.

There are other reasons to know about McCarthy. He invented Lisp (*list p*rocessing), the principal language of artificial intelligence (AI) and since its inception a fertile source of ideas for language design. As a teacher and a poser of puzzles, he has inspired other computer scientists in a variety of subspecialties from cryptography to planarity testing as the chapters on Rabin and Tarjan will illustrate.

Born in Boston in 1927 to Communist party activists, McCarthy lived in a family on the move. While his Irish Catholic father worked as a carpenter, fisherman, and union organizer, the family kept migrating, from Boston to New York and then to Los Angeles. His Lithuanian Jewish mother worked as a journalist for *The Federated Press* wire service then for a Communist newspaper and finally as a social worker. McCarthy connects his early interest in science with the political views of his family.

> There was a general confidence in technology as being simply good for humanity. I remember when I was a child reading a book called *100,000 Whys*—a popular Soviet technology book written by M. Ilin in the early 1930s. I don't recall seeing any American books of that character. I was very interested to read ten or fifteen years ago about some extremely precocious Chinese kid and it was remarked that he read *100,000 Whys*.

McCarthy remembers himself as an unremarkable teenager, but the evidence doesn't bear this out. As a high school junior, he got a copy of the course catalog from the California Institute of Technology and looked up the calculus books used for freshman and sophomore mathematics. He bought the books and worked out all the exercises. This enabled him to skip the first two years of college math when he finally arrived at Cal Tech in 1944.

In 1948, McCarthy began graduate studies in mathematics. That September, he attended the Hixon Symposium on Cerebral Mechanisms in Behavior at Cal Tech. The great mathematician and computer designer John von Neumann delivered a paper on self-replicating automata—machines that can create copies of themselves. Although no one at the conference explicitly related machine intelligence to human intelligence, von Neumann's talk sparked McCarthy's curiosity.

In 1949, having begun his Ph.D. work in mathematics at Princeton, McCarthy made his first attempt to model human intelligence on a machine.

> I considered an intelligent thing as a finite automaton connected to an environment that was a finite automaton. I made an appointment to see John von Neumann. He was encouraging. He said, "Write it up, write it up." But I didn't write it up because I didn't feel it was really good.

An *automaton* models a machine that moves from one state to another over time. For example, a standard transmission car moves from the "off" state to the "in-neutral-but-on" state when the driver engages the ignition. It then moves to the "in-first-gear-and-on" state when the driver switches gears to drive. An *interacting automaton* moves from state to state depending on its own state and on what it observes regarding the states of other machines. Some automata are intelligent (presumably, the driver is), but intelligence is not a necessary component. The interacting automata model attempts to establish a continuum between the two kinds.

McCarthy rejected his own first attempt at using automata to model human intelligence, but the idea of states and transitions would resurface in his work on situational calculus more than a decade later.

Throughout this time, McCarthy's interest in making a machine as intelligent as a human persisted. In the summer of 1952, a graduate student at Princeton—Jerry Rayna—suggested that McCarthy try to get some people who were interested in the topic of machine intelligence to collect papers on the subject. One of the first people McCarthy approached was Claude Shannon, inventor of what Shannon called a mathematical theory of communication. Shortened by others to *information theory,* Shannon's theory was first applied to telecommunications and later to linguistics, mathematics, and computer science.

> Shannon didn't like flashy terminology. He named the volume *Automata Studies.* And then the papers started coming in and I was disappointed. Not enough of them were about intelligence.
>
> So when I started to organize the Dartmouth project in 1955, I wanted to nail the flag to the mast and used the term *artificial intelligence* to make clear to the participants what we were talking about.

The Dartmouth 1956 Summer Research Project on Artificial Intelligence proved to be a landmark event in the history of computer science. The ambitious goal for the two-month ten-person study was to (in the words of the proposal) "Proceed on the basis of the conjecture that every aspect of learning or any other feature of intelligence can in principle be so precisely described that a machine can be made to simulate it."

The four organizers—McCarthy, Marvin Minsky (then at Harvard), Nat Rochester (a prominent IBM computer designer), and Shannon—made what in today's terms seems a quaintly modest request for financial backing from the Rockefeller Foundation: $1200 for each faculty-level participant and "railway fare for participants coming from a distance"—for a total of $7500.

In his part of the proposal, McCarthy wrote that he would study the relation of language to intelligence with the hope of programming a computer to "Play games well and do other tasks." Looking back at the conference almost forty years later, McCarthy describes, in characteristically blunt terms, his expectations.

> The goals that I had for that conference were entirely unrealistic. I thought that major projects could be undertaken during the course of a summer conference. The model was actually something that I had never attended but had heard of—a military summer conference on air defense.

While the conference came nowhere close to resolving the enormously difficult problem of creating a truly intelligent machine, it established goals and techniques that led to the recognition of artificial intelligence as a separate, and ultimately, a driving field of study within computer science. Although many of the conference attendees never pursued further research in the field, a few presented work of lasting impact.

Allen Newell, J. C. Shaw, and Herbert Simon of Carnegie Mellon University described a second version of their Information Processing Language (IPL 2). IPL2 resulted from the three scientists' effort to build a program, called the Logic Theory Machine, that could prove theorems in elementary logic and play games. To do this, they needed a programming language that could manipulate symbols for objects such as chess pieces and truth values of logical variables. Because this was very different from performing arithmetic on numbers, they suggested using what they called *list structure*.

To illustrate the use of lists for symbolic processing, let us take some liberties with *Alice in Wonderland*. Suppose a mutant Cheshire

Cat tells Alice "Either I am mad or the Hatter is mad." Suppose we let C, H, and A represent the respective assertions that the Cheshire Cat, Hatter, or Alice is mad. We might render the cat's statement in list format as (or C H). The Cat then tells Alice, "Either you or the Hatter is mad." A clever Alice might put this statement together with the other as (and (or C H) (or A H)). Finally, the cat says, "Only one of the three of us is mad." That is, at least two are NOT mad. Alice might represent this as (and (or C H) (or A H) (or (and (not A) (not C)) (and (not A) (not H)) (and (not C) (not H)))).

After putting these statements in list format, we can then form rules for list manipulation such as (and (or X Y) (or Z Y)) = (or (and X Z) Y). That is, if either X or Y and either Z or Y hold, then either X and Z hold or Y holds. Applying that rule and others will allow us to conclude (and H (not C) (not A)). According to the Cat, then, only the Hatter is mad.

The beauty of using lists for logical reasoning is that they can grow and shrink and reform themselves as the inferences proceed. Further, one can represent both rules and data in the same form. To most participants at the conference, list manipulation looked to be a clear winner. The other achievement of the Dartmouth Conference was Marvin Minsky's proposal to build a geometry theorem prover. Minsky had tried out some examples on paper and proposed that proving theorems in geometry might be a good application of the rule-based approach that Newell and Simon had advocated. Herbert Gelernter and Nathaniel Rochester at IBM decided to implement the program. Gelernter would later develop a tool to help organic chemists synthesize new chemicals. His son David is a renowned researcher and designer in parallel programming and in medical artificial intelligence. McCarthy acted as a consultant to the theorem proving project, giving him the chance to program intelligent behavior.

> Gelernter and his assistant Carl Gerberich took my suggestion to start with Fortran and made something which they called FLPL—Fortran List Processing Language. And they added some ideas of their own.

In 1956, John Backus and his IBM team had introduced Fortran, the first high-level programming language. Fortran freed programmers working on numerical computations from the difficulty of writing assembly language specific to each computer. To this day, it remains a lingua franca for scientific and engineering computing. FLPL was a first attempt to extend Fortran's capabilities to symbolic manipulation. In the summer of 1958, while working at IBM, McCarthy tried to use FLPL to write list programs for an application he knew extremely well from his high school work, the differentiation of algebraic expressions. The idea immediately demanded recursive conditional expressions,[1] and recursion was not possible in Fortran.

> If Fortran had allowed recursion, I would have gone ahead using
> FLPL. I even explored the question of how one might add recursion
> to Fortran. But it was much too kludgy.

As it turned out, IBM soon lost interest in artificial intelligence anyway. Some customers thought their jobs might be threatened by intelligent machines, so by the early 1960s IBM's marketing message was that computers were just dumb number-crunchers that would do what they were told—no more, no less.

Instead of tinkering with Fortran, McCarthy invented Lisp. Whereas Newell, Shaw, and Simon later described IPL as a language that grew complex over time, McCarthy describes Lisp as a language that became simpler over time.

Lisp stands for list processing language, and as might be expected all data in Lisp is represented as lists. These lists are contained within parentheses. For example, (Robert taught Dennis) might be a list representing the sentence "Robert taught Dennis." In this case, the order matters because it indicates who taught whom.

The list (cabbage lettuce strawberries) might represent a shopping list. In this case, the order doesn't matter—the three items can be bought in any sequence. In both examples, the list contains "atoms" as elements. An atom, unlike a list, has no component parts. A list may contain (and usually does contain) other lists as component parts. For example, (Robert splashed (Carol and Dennis)) reflects the grammatical

---

[1] For example, the derivative of y^2 results in $2y$ times the derivative of y; this is recursive because the derivative of an expression is defined in terms of the derivative of a component of the expression.

structure of the sentence in which the parentheses indicate that both Carol and Dennis are objects of the verb splashed. As a second example, (times 6 (plus x y)) would represent $6 \times (x + y)$. Here the order matters and the parentheses indicate that x and y are grouped.

In this way, lists can represent any of the standard mathematical structures that underlie science and engineering as well as sentence structures underlying language.

From the beginning, McCarthy had a team of eager, if captive, collaborators.

> When I returned to M.I.T. in the fall of 1958, Minsky and I had a big work room, a key punch, a secretary, two programmers, and six mathematics graduate students. We had asked Jerry Wiesner for these things the preceding spring in order to form an artificial intelligence project.
>
> We got them even though we hadn't prepared a written proposal. Fortunately, the Research Laboratory of Electronics at M.I.T. had just been given an open-ended Joint Services Contract by the U.S. Armed Forces and hadn't yet committed all the resources. I think that such flexibility was one of the reasons the U.S. started in AI ahead of other countries. The Newell-Simon work was also possible because of the flexible support the U.S. Air Force provided the Rand Corporation.

As the effort progressed, McCarthy tried to improve the language's expressive power. In 1959, while trying to show that the language could formulate any computable function, he added a feature that became known as *eval*.

Eval permits a program to define a new function or procedure and then execute it as part of that program. Most languages would force the program to stop and "recompile" before executing the new function. Since the *eval* function can take any function and execute it, *eval* plays the role of a "Universal Turing Machine," a universal simulator of other computers.

The *eval* notion serves very practical purposes. For example, a brokerage house runs its computing services twenty-four hours a day, seven days per week because of international financial market activity. Suppose someone writes a program to analyze Reuters stock

LISP

BASIC OPERATIONS IN LISP

Typical Lisp functions and procedures either take a list and break it apart or take a few lists and then form a new list. For example, the function "append," takes two lists and creates a new one by attaching one to the end of the other. This might be useful if you were to build a sentence out of noun and verb phrases. Let us look at the relationship between two characters, Bob and Alice: (append (Bob kissed) (Alice)) creates the list (Bob kissed Alice).

Another useful function, "reverse," might order the elements in a list from back to front. In that case (reverse (append (Bob kissed) (Alice))) would create (reverse (Bob kissed Alice)) which would then create (Alice kissed Bob). On the other hand (append (reverse (Bob kissed)) (Alice)) would create (append (kissed Bob) (Alice)) which would then create (kissed Bob Alice).

However Bob and Alice work out their relationship, we can see that a Lisp program can be composed from functions that take lists and produce new lists. This is what appealed to Backus when he invented FP.

RECURSION AND EVAL IN LISP

McCarthy made recursion a centerpiece of Lisp's processing strategy. Recursion is a technique for defining an operation in terms of itself. Programmers get around the taboo of circular definitions by making the defining instance refer to a simpler problem.

For example, if the list contains one element, you could define the reverse of a list as the list itself. Or if the list has several elements, you could define the reverse of the list as: append the reverse of all elements of the list *except the first* with the list containing the first element.

If that seemed like a mouthful, try this. Call the first element of list L the head of L, denoted (head L), and the rest of the elements the tail of L or (tail L). We then could write out this program in the spirit (though not the syntax) of Lisp as:

```
define (reverse L) as
if L has one element then L
else (append (reverse (tail L)) (list (head L)))
```

Let's try this on a concrete example—this time with a ménage à trois, Alice, Bob and Carol: (reverse (Alice Bob Carol)) = (append ((reverse (Bob Carol)) (Alice))) = (append ((Carol Bob) (Alice))) = (Carol Bob Alice).

Because functions and procedures are themselves defined as lists, they can be constructed through other functions and procedures. The eval function can then take such a function or procedure and execute it.

data in a new way. Brokers would want to use the program immediately, but only if they could do so without ever losing the use of their machines. *Eval* makes this possible.

The ideas embodied in Lisp appealed to the international committee charged with the design of Algol (the language for which Backus and Naur invented Backus-Naur form). In the Paris committee meeting on the language in 1960, McCarthy proposed both recursion and conditional expressions. After some haggling over notation, the committee accepted the idea.

Algol was the first language to adopt Lisp innovations, but certainly not the last. Descendants of Algol such as Pascal, C, Ada, and most other modern programming languages offer recursion and conditional expressions. Until recently, mainstream languages did not offer *eval*, mostly because language designers feared that giving programmers the ability to put new features into a running program was hazardous. But now that many programs must run twenty-four hours a day, seven days a week, this feature is more and more in demand, and most new experimental languages include something like *eval*.

Lisp has reigned as the standard language of artificial intelligence for almost forty years. McCarthy did not anticipate its longevity and even proposed changes to make it similar to Algol. AI programmers preferred Lisp's original syntax, however, and McCarthy, like Backus before him and Kay after, lost control over the language.

Making Common Sense Logical

Fortunately for his research ego, McCarthy had always viewed Lisp as a means to reach his major goal which was—and still is—"to make a machine that would be as intelligent as a human."

He elaborated on this goal in a paper entitled "Programs with Common Sense," published in 1959. That paper marked the starting point of his career-long quest to apply mathematical rigor to that elusive form of reasoning known as common sense.

To make his goal concrete, he defined a program with common sense as one which could "deduce for itself a sufficiently wide class of immediate consequences of anything it is told and what it already knows." As an example, he described the reasoning that leads a person sitting at his desk to use his car to go to the airport.

In the discussion following the presentation of the paper, the well-known logician and linguist Yehoshua Bar-Hillel characterized McCarthy's approach as "half-baked" and "pseudo-philosophical," saying that the reasoning McCarthy advocated could not be called deduction since the conclusion might not always hold. "Might it not be cheaper to call a taxi and have it take you over to the airport? Couldn't you decide to cancel your flight or to do a hundred other things?" Hillel asked.

In his reply to the criticism, McCarthy agreed that the paper rests on "unstated philosophical assumptions. . . . Whenever we program a computer to learn from experience, we build into the program a sort of epistemology." Working out that epistemology has been the focus of research for McCarthy and a handful of other AI theorists ever since.

Before you read about that work, you may wonder why anyone would want to build common sense into a computer program in the first place. What good would it do for science or society? McCarthy suggests an answer.

> All science as well as all specialized theories are embedded in common sense. When you want to improve these theories, then you go back to commonsense reasoning, which directs your experimentation.
>
> So if somebody wants to make a better chess program, then he makes experiments with the one he's got, giving it positions to analyze. All the reasoning about what experiments to make is in a commonsense framework.

You can find echoes of such sentiments in the writings of many scientists. For example, in his well-known lectures on physics, Richard Feynman made the following observation when discussing symmetry:

> All of our ideas in physics require a certain amount of common sense in their application; they are not purely mathematical or abstract ideas. We have to understand what we mean when we say that the phenomena are the same when we move the apparatus to

a new position. We mean that we move everything that we believe is relevant; if the phenomenon is not the same, we suggest that something relevant has not been moved, and we proceed to look for it.[2]

One of the primary virtues of commonsense reasoning is its resiliency. It adapts well when confronted with new facts about a situation. McCarthy gives the following example.

Imagine a traveler who wants to fly from Glasgow to Moscow via London. Now imagine that he has some facts about needing a ticket and about the effect of flying from one place to another.

Many programs can be made to reason that if he flew from Glasgow to London and London to Moscow, he'd be in Moscow. But what happens if he loses his ticket in London? You'll no longer be able to say that the original plan will work. But a plan that involves his buying another ticket should work. Not one of the existing applied programs allowed for this kind of elaboration of a situation's conditions.

In 1964, McCarthy, by then the head of Stanford's AI laboratory, suggested a type of logic called *situational calculus*, where a *situation* represents a state of the world. When an agent acts, a new situation results. What action the agent then takes depends on what he knows about the situation.

In McCarthy's airplane example, a traveler who doesn't realize that he has lost his ticket may go directly to the airport, but a traveler who realizes his loss may go to his travel bureau, where he'll deal with a (presumably intelligent) travel agent.

The situation calculus thus shares with the theory of finite automata the notion of transition. In the situation calculus, however, reasoning depends not only on the situation, but also on what the agent knows about the situation. The more he or she knows or can find out, the better the decision—one hopes. Situational calculus has appealed to many researchers, who have applied it or its variants in

[2]Richard P. Feynman, Robert B. Leighton, and Matthew Sands, *The Feynman Lectures on Physics* (New York: Addison-Wesley, 1977), vol. 1, p. 11-1.

many ways, but the theory brought into focus an enormous new problem.

In a world of many interacting agents, the situation relevant to one agent may change depending on what another agent does. In our commonsense world, however, we know that most of what other agents do shouldn't materially affect our decisions.

For example, the actions of our airplane traveler (agent A), who has lost his ticket, should not change if the president of the United States (agent B) buys a bran muffin on his morning jog in Washington, D.C. There is nothing in logic that tells us that this action is irrelevant, but our intuition leads us to this conclusion (unless we believe in unlikely conspiracies).

Writing with Patrick Hayes of the University of Edinburgh, McCarthy dubbed the general problem of compactly representing the many facts that remain unaffected by a particular action the *frame problem.*

> To put it simply, the frame problem is not having to mention all the things that don't change when a particular action occurs.

Most successful reconstructions of mysteries depend on insightful resolutions of the frame question. When the detective hero finds a solution, he identifies a connection among actions and agents that most people might have dismissed or overlooked altogether. Returning to *Silver Blaze,* for example, Holmes identified the culprit as the owner of the watchdog because the dog didn't bark. Few people, including his estimable sidekick, Dr. Watson, would have made this leap.

"It is one of those cases where the art of the reasoner should be used rather for the sifting of details than for the acquiring of fresh evidence," Holmes remarks at the outset of his investigation. "The tragedy has been so uncommon, so complete, and of such personal importance to so many people that we are suffering from a plethora of surmise, conjecture, and hypothesis. The difficulty is to detach the framework of fact—of absolute undeniable fact—from the embellishments of theorists and reporters."[3]

---

[3]*The Complete Sherlock Holmes,* p. 335.

For the computer to solve the frame problem in a setting where new and surprising facts may enter at any time, it must, like Sherlock Holmes, embark on a new kind of reasoning. In 1974, Marvin Minsky published *A Framework for Representing Knowledge* (M.I.T. Press) in which he argued that AI should not use logic, because logic is inherently too conservative.

To understand his argument, consider what happens if you hear that a friend has a bird named Banjo. Your image of the situation will probably include the belief that Banjo can fly. Now, if your friend tells you that Banjo is a penguin or has clipped wings or has had his feet tied or has a fear of flying, you will no longer believe that Banjo can fly. Without contradicting the fact that Banjo is a bird, the new information causes you to retract your initial conclusions.

Minsky dubbed this form of reasoning *nonmonotonic.* Classical deduction is always monotonic—new information never causes you to retract your original conclusions provided the new information doesn't contradict the old. Nonmonotonicity pervades everyday reasoning. We believe that the President's wolfing down a muffin in Washington, D.C. won't affect the traveler at Heathrow, but our beliefs will change if we learn that his eating the muffin signaled the start of a saxophone concert at Heathrow.

McCarthy viewed nonmonotonicity as a way to resolve the frame problem. Intuitively, an intelligent agent should reason as if distant acts wouldn't matter, but may have to reason anew upon discovering that those acts may in fact be relevant. McCarthy, unlike Minsky, felt that logic constituted the only solid foundation for resolving the problem. To do so, logic needed a formal basis for conjecture.

McCarthy proposed *circumscription,* a rule that permits an agent to conjecture as follows: "I already know about all objects that are likely to have a property P. So if I come upon a new object, then I will assume that it doesn't have property P."

In the case of Banjo, for example, you assume that Banjo flies. That is, you circumscribe the nonflying property for birds. In contrast, in the case of elephants, you circumscribe the *flying* property, assuming, in the absence of evidence to the contrary, that Jumbo the elephant cannot fly—it lacks the flying property. Thus, circumscription is a method to make educated guesses about things. It may prove wrong when you learn more, but in the meantime it allows you to form useful hypotheses.

To know which properties to circumscribe in a given situation, you must know facts about the world. In the cases of Banjo and Jumbo, you must know that normal birds fly whereas normal elephants don't. People apply such knowledge almost without thinking, but the issue presents a fundamental challenge.

> For example, the encyclopedia will tell you Napoleon died in 1821 and Wellington died in 1852 and that when he died, Napoleon was the British government's prisoner on the island of Saint Helena.
>
> But the encyclopedia itself would only accidentally tell you that Wellington heard of Napoleon's death—you have to apply common sense to encyclopedic knowledge in order to do it.

You might reason as follows: Wellington probably did hear of Napoleon's demise, because he had been obsessed by Napoleon for most of his professional life. Thus you could circumscribe "not hearing about the deaths of people you obsess about." In the case of the simpler question—did Napoleon hear of Wellington's death?—you might circumscribe "hearing of deaths of people who die after you die."

The fundamental issue is how to use the context of a situation to make natural conjectures, such as that generals care about their opponents. McCarthy has started to develop a new logic form manipulating and using contexts.

> Part of the goal is simply to understand what people are saying. If your AI system can't use contexts, then it will not be able to interpret the statements that people make.

To appreciate the magnitude of this problem, consider the fact that applying the correct context is not only difficult for machines; people sometimes have trouble, too. The late Speaker of the House Thomas P. (Tip) O'Neill used to tell the story of welcoming the newly elected President Ronald Reagan to his office. O'Neill told Reagan that he had Grover Cleveland's desk. Reagan smiled and said that he had once played Cleveland in the movies. Reagan was talking about

Grover Cleveland Alexander, the baseball player, while O'Neill was referring to the two-term President of the United States.

So Many Questions, So Few Solutions

Opening the Pandora's box of commonsense reasoning led McCarthy to study formalisms that allow the addition of arbitrary new facts—what McCarthy called *elaboration tolerance*. Once elaboration is allowed, new facts may force a reasoner to retract some previously drawn conclusions while retaining others. Reaching both those correct and incorrect conclusions may require circumscribed conjectures based on the prior likelihood of certain predicates, e.g., birds fly but elephants don't. Figuring out which conjectures are likely to hold requires knowledge and a way of organizing knowledge into contexts or microtheories. Knowledge evolves as the reasoner acts and therefore creates new facts. Each capability depends on the other.

In McCarthy's 1959 paper, he observed that the goal is to create a machine that can "perform certain elementary verbal reasoning processes so simple that they can be carried out by any non-feeble-minded human." Put in those terms, the task sounds eminently doable. Yet the goal remains distant.

So, you might ask, were McCarthy's many attempts at providing a logical foundation for artificial intelligence a success or a failure? By his own reckoning, little of his work since situational calculus has been put in practice. The builder of an expert system for medical diagnosis or stock price prediction, for example, will go to great pains to avoid putting nonmonotonic reasoning or general context inference into his system. The difficulties uncovered by McCarthy and his colleagues serve as "keep off" warnings.

But the future may be quite different. If a project is broad as well as deep—like Doug Lenat's Cyc, which tries to encode tens of millions of facts and rules—it faces many unpleasant difficulties. Any such project that uses logic must build upon McCarthy's work. (As we will discuss in the Lenat chapter, Lenat's project uses "microtheories" developed by R. V. Guha, who was a student of McCarthy's.) Will the current logical tools work? McCarthy doesn't claim to know.

Progress in using logic to express facts about the world has always been slow. Aristotle invented no formalisms. [Gottfried] Leibniz never invented propositional calculus even though that is an easier formalism than the infinitesimal calculus which he co-invented with Newton. [George] Boole invented propositional calculus, but never invented predicate calculus. [Gottlob] Frege invented predicate calculus but never tried to formalize nonmonotonic reasoning. I think that we humans find it difficult to formulate many facts about our thought processes that are apparent when suggested.

Alan C. Kay

A CLEAR ROMANTIC VISION

All understanding begins with our not accepting the world as it appears.

—ALAN C. KAY

cientists have long dreamed of machines that would improve the quality of our lives. In the seventeenth century, the philosopher and mathematician Gottfried Wilhelm von Leibniz imagined a machine that could settle all disputes through logical reasoning. Three centuries later, in 1945, Vannevar Bush, the inventor of a proto-computer called the differential analyzer, published an article in the *Atlantic Monthly* entitled "As We May Think." In it, Bush speculated about a time in the future when a "memex" device would enable the average person to browse through a microfiche library. By following links from one fiche to another, the individual would gain a cross-disciplinary view of any subject. Robert Heinlein, the science fiction writer, later used this idea in a short story. It intrigued a 14-year-old boy named Alan Kay.

Kay has since come up his own idealistic contraptions. His work on the FLEX machine and subsequent design of the Dynabook prototype and the programming language Smalltalk have enormously influenced the design of today's personal computers. *Object-orientation*, a term Kay coined, has become the primary paradigm for programming in the middle 1990s. His current collaboration with a Los Angeles public school reflects an approach to education antithetical to the one Kay received—and despised.

Born in 1940 in Springfield, Massachusetts, Kay moved with his family a year later to Australia, his father's birthplace. Kay's father, a physiologist, designed prostheses for arms and legs; his mother was an artist and musician.

> Since my father was a scientist and my mother was an artist, the atmosphere during my early years was full of many kinds of ideas and ways to express them. I did not distinguish between "art" and "science" and still don't.
>
> My maternal grandmother was a schoolteacher, suffragette, lecturer, and one of the founders of UMASS, Amherst. My maternal grandfather was Clifton Johnson, a fairly well-known illustrator, photographer, and writer (100+ books). He was also a musician, and played piano and pipe organ. He died the year I was born, and the family myth is that I am the descendant most like him, both in interests and in temperament.

In early 1945, when a Japanese invasion of Australia appeared likely, the Kay family returned to the United States. From 1945 to 1949, they lived in the Johnson farmhouse outside of Hadley, Massachusetts. Kay had learned to read in Australia at the age of three and reveled in the new surroundings—nearly 6000 books in the house and many drawings and illustrations.

> One book I read was called *Rockets, Missiles, and Space Travel* by Willi Ley. The thing that struck me was that when you go from one planet to another, you wouldn't go the way you thought you would. You don't aim the rocket ship at the planet, you aim the rocket ship at where the planet is going to be.

Although visions of space travel excited Kay, school didn't.

> By the time I got to school, I had already read a couple of hundred books. I knew in first grade that they were lying to me because I had already been exposed to other points of view. School is basically about one point of view—the one the teacher has or the textbooks

have. They don't like the idea of having different points of view so it was a battle. Of course, I would pipe up with my five-year-old voice.

Kay learned music from his musician mother. He was a boy soprano and soloist in the school choir. As a teenager he played guitar and in his twenties was good enough to earn money as a professional. He sees a direct relationship between music and computation.

> Computer programming is a bit like a Gregorian chant—a one-line melody changing state within larger scale sections. Parallel programming is more like polyphony.

In the terms of Kay's analogy, a section of a Gregorian chant contains many variations upon a theme. Within a loop of a computer program, the same sequence of instructions is repeated many times. Each time, the sequence starts with a different value, until a "stopping value" determines that the loop is at an end and it is time to go on to the next loop, or in the case of Gregorian chant, the next melody. In polyphony, many different themes are played at the same time just as in parallel programming many different sequences of instructions are executed at the same time.

When Kay's father got work in a New York hospital in 1949, the family moved to Long Island and Kay attended Brooklyn Technical High School, which with Bronx Science ranked as one of the best science high schools in New York City. His insubordinate behavior resulted in his suspension. Soon after, he came down with rheumatic fever. He thought he might have to repeat his senior year, but he had already accumulated more than enough credits to graduate from high school. The question was, what next?

> I went to a school in West Virginia called Bethany College and I wound up majoring in biology with a minor in mathematics. I wound up also getting thrown out for protesting the Jewish quota in 1961. So I went to Denver to teach guitar for a year. Then the Army was interested in drafting me. There was a provision if the army was going to draft you, you could apply for one of the volunteer services.

> So I wound up spending a couple of years in the Air Force. Being a reader, I read all of the Air Force regulations and discovered—I was training for an officer—if I got out of officer training I'd be an enlisted man on the balance of a two year enlistment instead of a four year enlistment. So I got out of that.

While in the Air Force, Kay passed an aptitude test to become a programmer and worked with the IBM 1401, an extremely popular machine for its time, with about 15,000 out in the market.

> They needed programmers. This was back in the days when programming was a low status profession and most of the programmers were women. My boss was a woman. They also were taking linguists. . . . It was actually a pretty interesting bunch.
>
> I had a friend who was a black guy who did what today we would call an operating system [the control program of a computer]. You have to realize that the 1401 had 8K [8000 characters worth] of memory so the operating system had to be less than 1K. He had done this wonderful thing that allowed the machine to do really complex batch processing jobs and it was wonderful. I helped him work and got into the style of programming.

The Germ of a New Idea

The ability to start with an idea and see it through to a correct and efficient program is one prerequisite for a great software designer. A second is to see the value of other people's good programming ideas.

In 1961 Kay worked on the problem of transporting data files and procedures from one Air Force air training installation to another and discovered that some unknown programmer had figured out a clever method of doing the job. The idea was to send the data bundled along with its procedures, so that a program at the new installation could use the procedures directly, even without knowing the format of the data files. The procedures themselves could find the information they needed from the data files. The idea that a program could use procedures without knowing how the data was represented struck Kay as a good one. It formed the basis for his later ideas about objects.

After leaving the Air Force, Kay decided it was time to finish college, so he matriculated at the University of Colorado.

> Colorado had a better mathematics department than Bethany but a worse biology department. I spent most of my time in Boulder in the theater, writing stage music. They had a theater that was open to everybody, even nonmajors. They had something like 25 productions a year—it was fantastic. I wrote one based on *The Hobbit* and got involved in some productions.

Though he seriously considered pursuing a musical career, Kay graduated from Colorado in 1966 with a double major in mathematics and molecular biology. He was again faced with a career decision.

> I thought about medicine but I didn't feel I was responsible enough. I still feel that way today.

He also considered philosophy, but decided against it. He finally decided to try computer science at the University of Utah.

> All I knew was that Utah was above 4000 feet in altitude and it had a Ph.D. program. I loved the mountain air. I had thought about going to Wisconsin in philosophy but fortunately that didn't turn out. Boulder didn't have a Ph.D. program. So I just showed up at Utah literally with a dime in my pocket. Computing was something I could do.

There were six other graduate students and three professors in the computer science department, then headed by Dave Evans, who had previously been in charge of computer science at the University of California at Berkeley.

> At Utah before you got a desk you got a stack of manuscripts and you had to read the stack. It described Sketchpad. Basically you had to understand that before you were a real person at Utah.
>
> They also had a tradition there that the latest graduate student got the latest dirty task to do. Mine happened to be on my desk— a pile of tapes and a note which said "This is the Algol for the UNIVAC 108; if it doesn't work, make it work." It turned out to be the first Simula.

Ivan Sutherland's Sketchpad, introduced in 1962, was the first system ever built for interactive computer graphics and it was remarkably sophisticated. It had a notion of master drawings and instance drawings. A programmer could define various constraints on the parts of a *master drawing*. These constraints could be simple—the end of line L should be 1 inch from some point P—or exceedingly complex—this truss of mass M should bend as if Newton's laws applied to it. Sketchpad would take a drawing (an instance), check it against the constraints, then attempt to change the drawing to conform to the constraints.

Simula, developed in 1965 by Kristen Nygaard and Ole-Johan Dahl in Norway, supported a similar distinction between masters and instances, although it used different terms. In both languages, the programmer would define behavior in the master and then each instance would conform to that behavior. Kay thought a lot about these ideas. He was looking for a single basic building block that would permit a simple, powerful style of programming.

> The big flash was to see this as biological cells. I'm not sure where that flash came from but it didn't happen when I looked at Sketchpad. Simula didn't send messages either.

The biological analogy suggested three principles to Kay. First, every cell "instance" conforms to certain basic "master" behaviors. Second, cells are autonomous and communicate with one another using chemical messages that leave one protective membrane and enter through another one. Third, cells can differentiate—the same cell can, depending on context, become a nose, eye, or toenail cell. Kay would include the master-instance distinction, message passing, and differentiation later in his design of Smalltalk, but for now these were just ideas that seemed important but had no particular home.

FLEX and Dynabook

Financing for many of Utah's projects came from the Department of Defense's Advanced Research Projects Agency (ARPA) during what many look back at as ARPA's golden age. J. C. R. Licklider, who ran the agency from 1962 to 1964, had a background in psychology and was very much interested in the human-machine interface.

From 1962 to 1970, ARPA funded a number of risky projects, some of them nonmilitary, just because the agency had faith in the people who were running them. In 1970 the Mansfield Amendment required ARPA to fund work concerned with military aims only. The agency then changed its name to DARPA (Defense Advanced Research Projects Agency). In 1993 the federal government changed the name back to ARPA, reflecting a renewed civilian focus.

In 1963, with ARPA funding, Wesley Clark of M.I.T.'s Lincoln Laboratory developed the first personal computer, LINC. Weighing in at several hundred pounds, it bore little resemblance to today's palm-tops, but the mere existence of such a project created excitement. Similar efforts soon followed and Dave Evans heard about them.

> Evans had been vice president of Bendix—they were doing real world things. One of the things he would often do was get his students consulting jobs. A couple months after I was there he took me to a local consulting job.
>
> Dave Evans knew this guy that was doing desktop machines in 1967, but didn't know about software. So since I knew about software, it turned out to be a perfect match.

The "guy" turned out to be Ed Cheadle. The project turned out to be the FLEX machine.

> Cheadle was a wonderful guy, a big Texas jovial guy. The two of us hit it off really well and started talking about what the machine really should be. He wanted it to be an engineering workstation that would work for doctors, lawyers, and professionals.
>
> Of course right away you get into the basic problem with personal computers—once you extend the franchise to everyone, you have a problem—you can't anticipate the needs. So I very early on got interested in extensible languages [computer languages that a user can tailor to the terminology and operations in his or her field].
>
> The Rand Corporation had experimented with JOSS, a system designed for noncomputer users—in this case economists.
>
> The idea was that FLEX was supposed to be like JOSS. It pretty much was. It had multiple windows and it even had something a little bit like icons.

Although it had some features of modern personal computers, Kay felt uneasy about FLEX, partly because of its bulky 350 pounds. In the summer of 1968, he gave a talk on the FLEX machine at the University of Illinois. The talk was well received, but Kay remembers best the tour that followed.

> We saw the first flat panel display. We came back from the conference and I spent half the night using Moore's law trying to figure out when I could get the FLEX machine on the back of a small display.

In 1965 the physicist Gordon Moore had predicted that integrated circuits on chips would double in density every year starting from the base year of 1959. That meant that the size of circuitry would be cut in half every year. Calculating on that basis, Kay figured that FLEX's 20,000 circuit elements would be able to fit behind a screen by 1980—not a bad guess as it turns out.

In the fall of 1968, Kay visited Seymour Papert at M.I.T.'s Artificial Intelligence Laboratory. Papert, who had studied with the Swiss child psychologist Jean Piaget, pioneered methods of teaching children to program for themselves in a simple programming language, LOGO, that Papert himself had developed. Kay saw public school kids in Lexington, Massachusetts, doing real programming. That experience changed his mind about personal computers.

> One of the metaphors we used for personal computers up to that point was that it was a car as opposed to an institutional railroad.
>
> When I saw Papert's stuff, that metaphor didn't hold up. You don't worry about kids driving. But you sure worry about kids learning the media and the symbol systems of civilization.

After completing his Ph.D. at Utah, Kay spent a year at Stanford's Artificial Intelligence Laboratory in 1969–1970, where he taught systems design courses.

> I really wasn't interested in AI. Part of the problem was that it was too hard. My biological background meant that it was very hard to set up a viable standard for AI. Every time I sat down and thought about it—what they were satisfied with was just not satisfying. They were doing stuff that didn't have a particular relevance to what I would call intelligence.

The one exception was Marvin Minsky as he evolved this "society of minds" idea. Minsky was always impressed with biology and the more he thought about it, the more biologically oriented he got.

While teaching at Stanford, Kay began ruminating about another type of personal computer—a book-sized machine that could link the user, especially children, to the world.

In July 1970, Xerox's chief scientist Jack Goldman persuaded the company's senior management to establish a long-term research center in Palo Alto, California. Xerox hired Bob Taylor, who had been a visionary at ARPA, to run the Palo Alto Research Center, or PARC as it came to be called. Taylor offered Kay the opportunity to "follow his instincts." So Kay started working on a design for a notebook computer called KiddiKomp which had a small (14 inch by 14 inch) television screen as the display, a detachable keyboard, and a mouse.

Soon after hiring Kay, Taylor nabbed most of the people from a small, faltering company called the Berkeley Computer Corporation. The group included Butler Lampson (who went on to win the Turing Award in 1993) and other stars. Then came many researchers from Doug Engelbart's pioneering user interface group at the Augmentation Research Center of the Stanford Research Institute, including Bill English, coinventer of the mouse. English encouraged Kay to form a group and write out a budget.

A reluctant manager, Kay formed the Learning Research Group (LRG) and hired people who shared his enthusiasm about the notebook computer idea. By the summer of 1971, Kay had begun work on a new language called Smalltalk.

The name was so innocuous that if it did anything, people would be pleasantly surprised.

The Invention of Object Orientation

Smalltalk was true to its biological analogue: autonomous cells communicating with one another through messages. Each message contained data, a sender return address, a receiver address, and the operation the receiver was to perform on the data. Kay wanted this simple message mechanism to apply throughout the language. By

September 1972 he had simplified the basic ideas so that a complete definition of Smalltalk could fit on one page. These ideas formed the kernel of what Kay called *object orientation,* a principal software technology of the 1990s.

Overuse has almost caused the term "object orientation" to lose its meaning. Its fundamental power, however, springs from its metaphorical relationship to autonomous, communicating biological cells. Just as cells in specific surroundings produce certain proteins in response to certain chemical messages, computational objects will produce certain responses to computational messages. For example, a computational object that represents a video game will respond to messages such as clicks on buttons and mouse movements by changing its display and adjusting the score. The player needs to know only about the external behavior of the game, not how it goes about its work. Hiding information about internal activity is the essence of object orientation.

Since all communication is through messages, an object can be lifted from one context and placed in another, provided the messages it receives and responses it gives are appropriate to the new context. This implies that software objects can find uses in places unimagined by the object designer, much in the same way that a cleverly designed car radio will one day be installed into cars that aren't yet on the drawing board.

What good is this? Software designers can now, in principle, build programs the way civil engineers design buildings: they can purchase common components from foundries and attach them together with customized software rivets. For example, a software foundry might define the behavior for image objects, allowing them to be rotated, expanded, projected, recolored, stretched, and so on. Any programmer who needs to include images can order the image "class" from the factory and add customized behaviors as needed. Further, just as a prefabricated trailer can be built from component sinks, toilets, desks, and so on, new behaviors can be built from component ones in an object-oriented system. Thus, a database of images offering search facilities as well as image functions can be built from database and image classes. Thanks to Kay's invention, the world no longer needs a new language for each application, since programmers can create objects to mimic any ontology.

Smalltalk: Kids as Designers

In September 1972, Kay finished the first design of Smalltalk. Within a few days, Dan Ingalls had the language running. The next problem was to get children to learn and use Smalltalk. Kay's goal was ambitious: he wanted Smalltalk to revolutionize the way children were educated. Influenced by Piaget, Jerome Bruner, and Rudolph Arnheim, Kay had decided that kids would learn more through images than by reading straight text. Using Smalltalk, Ted Kaehler and Diana Merry provided basic graphics and Bob Shur built an animation system (at two to three frames a second), all in short order.

Kay recruited Adele Goldberg and Steve Weyer, who were at Stanford, to work with the kids. In 1974 Goldberg devised a method to teach Smalltalk through a little graphic box called "joe." A command like "joe turn 30!" would rotate the square by 30 degrees; "joe grow 15!" would make joe grow bigger by 15 percent. The little box could be replicated (e.g., to jill), could be turned continuously (using "forever do"), and so on.

Some of the children caught on quickly. A 12-year-old designed a drawing system much like Apple's MacDraw (or Windows' Draw). A 15-year-old created a program to design circuits. Such superachievers comprised only 5 percent of the children Goldberg taught, according to Kay. Most of the others enjoyed working with joe, but they were not instant hackers. However, the fact that anyone was able to do anything so quickly was a good sign.

But Xerox management was not as enthusiastic as the kids. In 1976 bureaucratic inertia triumphed. Management decided against committing manufacturing resources to PARC's hardware or software designs. In rebuttal, Kay and his colleagues argued that the company should risk more not less. In one memo, Kay (correctly) predicted that there would be millions of personal computers in the 1990s and that these would often be hooked to global information utilities (like the Internet). Kay argued that Xerox had the manufacturing resources, the marketing base, and an "incredibly large percentage of the best software designers in the world" necessary to dominate this market. It didn't happen.

> They had no sense of the significance of what was there—there was no tradition. We didn't do a good job preparing them.

Even if we had done a good job of preparing them, business-men are more like engineers than like scientists. They basically want to do more than understand. At some point with new phe-nomena, you have to spend some time understanding it, not just worrying about what are we going to do with it. It happened to Xerox itself in the fifties when Arthur D. Little [the consulting firm] predicted there would never be a market for xerography.

Although Xerox never pursued PARC technology with much en-ergy, Apple certainly did. In 1979, Steve Jobs, Jeff Raskin, and other technical people from Apple came to Kay's group for a demo. They were impressed with what they saw, but they were even more im-pressed when Jobs complained about the screen display algorithm and Dan Ingalls rectified the problem in less than a minute. Jobs tried to buy the software from Xerox, but Xerox, though a minority invester in Apple, refused to sell it.

Smalltalk, now marketed by a Xerox offshoot called ParcPlace, continues to win adherents because of its incredible plasticity. A few years ago, Texas Instruments had to decide whether to program a project using the currently more popular language C++ or using Smalltalk. The Smalltalk team at TI proposed a "code-off." Three of them would go against any number of C++ programmers in an ex-periment: beginning at an agreed time in the morning, someone would specify a problem to be programmed by both teams; at noon, an aspect of the specification would be changed. The contest was to determine which team could achieve a more functional program. At Texas Instruments anyway, the Smalltalk team won easily.

Turning Schools Upside Down

Kay spent a year at Atari before becoming an Apple Fellow in 1984, where he currently serves as a kind of philosopher-in-residence. Among the projects he has supervised are Vivarium, originally con-ceived when he was at Atari, now a joint venture with M.I.T.'s Media Lab and the Center for Individualization (a magnet public school) in Los Angeles.

Vivarium offers 8-year-old students the opportunity to use com-puters to record real data and compare results with children at other schools. For example, students at the Center for Individualization have compared automobile traffic in their community with that of other communities of similarly equipped schools worldwide. Media lab software permits children to simulate ecosystems and test questions like, "Should an animal that is really hungry try to eat its predator?"

Superficially, one could argue that Kay's prototype for personal computing, the Dynabook, has arrived. After all, we have laptops, modems, handwriting recognition systems, database access, and hypertext. But Kay remains troubled by the potential for computers turning into "some kind of mass opiate."

> People rarely want what they need. Their wants are there for one reason and their needs for another. It is important for technology to match up needs and wants.
>
> If you put computers out there without value systems—that's like putting pianos out without compositions—you get a chopsticks culture [the melody, not the eating utensils]. You should make kids media guerrillas.
>
> Socrates complained about writing. He felt it forced one to follow an argument rather than participate in it. Computers can be far more animate than books. It should be possible for every kid everywhere to test what he or she is being told either against the arguments of others or by appeal to computer simulation. The question is: will society nurture that potential or suppress it?

ᴀLGORITHMISTS

HOW TO SOLVE PROBLEMS FAST

You know what algorithms are if you have ever followed a recipe, for recipes are culinary algorithms. Julia Child is someone who creates culinary algorithms for a living, and each of them must tell the amateur cook precisely what to do in readily understandable terms. Her recipe for coq au vin or charlotte russe must satisfy the chef's goals: good-tasting food cooked within a reasonable amount of time.

Like a world-class chef, a computer can follow an almost infinite number of recipes to solve an ever-increasing number of mathematical problems from cracking codes to drawing cartoons. The job of the algorithm designer is to find a recipe that solves these problems as fast as possible.

A good algorithm can—and every computer practitioner has seen this—shorten the time it takes to solve a problem from days to seconds. To see how such an improvement could be possible, consider your method for looking up an entry in a telephone book, say that of the "Egg Electric Company." You open to the first third of the book. If you've arrived at a letter past E, you flip backward; if before E, you flip forward; if E, you look at the second letter and make your decision on that basis. You estimate where you think the word will be, flip back if you've gone too far and continue forward if you haven't gone far enough. You zero in on the page after at most eight or so such glances. That is a much, much better algorithm than starting at ACME and reading down from there.

Finding good algorithms is what made the scientists in Part II of this book famous. Programs embodying their algorithms run trillions of times a day on your network, inside your spreadsheets, and in your word processor. In addition to inventing fundamental algorithms, however, these individuals have instilled algorithmic discovery in computer science with a three-part aesthetic of simplicity, correctness, and efficiency:

- An algorithm should emphasize the handful of ideas that make it superior to previous approaches to the problem it solves.
- The algorithm's proof should be built around the interaction of those ideas.
- The algorithm's time requirements should be easy to determine even as the size of the problem arbitrarily increases.

Edsger W. Dijkstra began his professional life shortly after World War II, when people thought of computers as marvels of engineering and tools for physicists, but unworthy of serious mathematical attention in their own right. It took him many years to publish his algorithm for finding the shortest path from one place to another, partly because mathematicians preoccupied with the problems of infinities could not see the value of a fast solution to a problem that anyone could solve in some finite time.

Later, when Dijkstra turned to the problem of preventing two programs from damaging one another other by colliding on common data—one of the fundamental issues in software design—he had to justify the importance of that problem by pointing to the engineering challenge of controlling the interaction between a keyboard and a computer. His later research has concerned techniques for *proving* that programs work correctly, an unpopular approach even today. A scientist without regard for fashions in research, Dijkstra has managed to shape his field by force of good taste and self-confidence.

Michael O. Rabin has played a central role in algorithmic development over the past thirty-five years. His early work with Dana Scott is the basis for computer language processing. His later work on randomized algorithms (ones that have a tiny probability of making a mistake but are very efficient) has given rise to a huge outpouring of developments in cryptography and network computing. Rabin summarizes the technique of randomized algorithms this way: "Flip a coin and use it wisely."

Donald E. Knuth brought mathematical precision to algorithms. For example, given five ways to solve a problem, Donald Knuth has told us which one is best and in which cases. A man of Herculean work habits, he has writen a classic multivolume textbook about every important result in all of computer science. His own inventions include an algorithm called LR(k) parsing that enables modern language translators to convert high-level programming languages to machine language.

Leslie Lamport spent much of his young adult life studying physics and thinking about special relativity and the ordering of events in space/time. This understanding gave him an unusual point of view when people started discussing distributed systems (collections of multiple processors linked together by a network). Lamport suggested that events in a distributed system should be thought of as having only local temporal order, since any notion of global time depended on keeping clocks perfectly synchronized or sending a message to a clock server. He then offered an algorithm for approximate clock synchronization. His model and his algorithms have laid the foundations for much of the infrastructure for cyberspace.

Robert E. Tarjan has always enjoyed drawing circles and lines between the circles. In the process, he has invented algorithms to lay out road networks on a surface, find the best flows through networks, and store historical snapshots. He has also brought elegance to the analysis of algorithms. Tarjan has given the world precise new criteria for judging the quality of algorithms, drawing upon such unlikely concepts as amortization in investing and competitiveness in economics.

A lost traveler in rural Maine might stop for directions to his final destination only to be told, "You can't get there from here." The traveler might then refer to his map to solve the problem, because he intuitively knows there is a way. Computer scientists find themselves in a similar situation but without the traveler's certainty. *Stephen Cook* and *Leonid Levin* defined a family of problems that include many interesting questions in circuit design, logic, resource allocation, and scheduling. They showed that either all of these problems are hard or all are easy. The trouble is, they don't know which. Neither does anyone else. Imagine Julia Child describing a meal that she does not know how to cook—or even whether it can be cooked at all.

Edsger W. Dijkstra

APPALLING PROSE AND
THE SHORTEST PATH

*I asked my mother [a mathematician] whether mathematics was a difficult topic.
She said to be sure to learn all the formulas and be sure you know them.
The second thing to remember is if you need more than five lines to prove
something, then you're on the wrong track.*

—EDSGER W. DIJKSTRA

cience, like dress design, has its fashions. The few
who ignore fashions in order to grapple with the
fundamental questions of their discipline take a big
gamble. Those who succeed earn the right to criticize. Throughout a
career dating back to the 1950s, Edsger W. Dijkstra has both gambled
successfully and criticized severely. His work on the shortest path algorithm and mutual exclusion is characterized by an Old World elegance and simplicity that he would like the rest of the world to share.

Born in Rotterdam in 1930, Dijkstra is the son of two scientists—his father was a chemist; his mother, a mathematician. Early
on, Dijkstra demonstrated an aptitude and a liking for science.

My older sister had a Meccano [like an American Erector set]—
long metal strips with holes. I made a lot of machines. I remember

I constructed two special cranes. One was constructed in such a fashion that no matter what its load was, its center of gravity was right above the fulcrum. The other one was such that when the distance between the center of the crane and the load was changed, the load remained the same height.

At age 12, in 1942, Dijkstra entered the Gymnasium Erasminium, an elite high school, where he received a traditional Dutch education— classical Greek and Latin, French, German, English, biology, mathematics, physics, and chemistry. The war brought hardship to most Dutch civilians including Dijkstra and his family. Toward the end of the occupation, when food was scarce, his family sent him out of the city.

I traveled with a friend of a friend of my father's who still had a car. We drove to the country. There was no gasoline. The car had to run on methane. . . . The radiator broke. It was freezing. I was fourteen and very weak—my heart couldn't manage more than forty beats per minute.

The young Dijkstra rejoined his family in July 1945. Political idealism was in the air. Dijkstra thought he might study law and serve his country in the United Nations, but his father dissuaded him.

I was talked out of law—the grades I had on my final examinations for mathematics, chemistry, and physics were so glorious.

Instead, he entered the University of Leiden where he had to choose between physics and mathematics.

I decided that if I didn't study physics at the university, I would never do it. I felt that mathematics would look after itself.

Having elected to study theoretical physics, Dijkstra observed that many problems in the field required extensive calculation, so he decided to learn to program. Thanks to their wartime code-breaking work, the British led the development of European computing in the 1940s and 50s. In 1951, Dijkstra attended summer school in programming at Cambridge University. In March 1952 he got a part-time job at the Mathematical Centre in Amsterdam, where he became progressively more involved in computer programming.

The Mathematical Centre was housed in an old school. The machine, called the ARMAC, occupied a classroom. It had a magnetic drum for memory [a rotating magnetic cylinder having recording heads capable of reading and writing from the outside surface]—advanced for the standards of the day.

In the early 1950s, before the advent of Fortran or Lisp, programmers wrote to suit the idiosyncratic design of each computer. A typical programmer would receive a list of the instructions that the machine could perform. If the hardware designers could simplify their design at the the expense of programming complexity, they would not hesitate to do so. The hardware designers working with Dijkstra, however, did just the opposite.

> They would never include something in the machine unless I thought it was okay. I was to write down the functional specification that was the machine's reference manual. They referred to it as "The Appalling Prose"—it was as rigorous as a legal document.

Dijkstra still had not committed himself to a career in programming—a field that was virtually unknown in the Netherlands.

> I finished my studies at Leiden as quickly as possible. Physics was a very respectable intellectual discipline. I explained to [Adriaan] van Wijngaarden [his advisor and an early Dutch computer pioneer] my hesitation about being a programmer. I told him I missed the underlying intellectual discipline of physics. He agreed that until that moment there was not much of a programming discipline, but then he went on to explain that automatic computers were here to stay, that we were just at the beginning and could not I be one of the persons called to make programming a respectable discipline in the years to come?

Dijkstra's doubts reflected the widespread ignorance about programming. When Dijkstra applied for a license to marry M. C. Debets, a colleague who was also a programmer, the bureaucrats did not recognize programmer as a profession, so he reluctantly labeled himself a "theoretical physicist."

Shortest Paths

Still at the Mathematical Centre, Dijkstra was asked to demonstrate the powers of the ARMAC for the forthcoming International Mathematical Conference of 1956. He started to think about the problem of determining the shortest route between two points on a railroad map. One sunny Saturday morning Dijkstra and his wife were sitting on the terrace of a cafe sipping coffee. Suddenly he fell silent.

> I was engaged in thought. My wife knew such periods. . . . The problem was so simple that you could find the solution without pencil and paper.

Figure 1 represents diagramatically—using highways instead of rail lines—the problem Dijkstra posed to himself. The task is to find the fastest route from City S to City T—the shortest path. Try the problem before looking at the legend text or the next paragraph.

Dijkstra's basic approach is to form an ever-growing "core set" of cities between your origin, City S, and your destination City T. At any given step of the algorithm, you know the minimum time to drive to every city in the core set. Initially, the core set consists of City S and it takes no time to drive there. At each subsequent step, you find a city outside the core set, call it X, having the property that the time to drive from City S to X is shorter than the time to drive to any other city outside the core set. Since it takes at least 0 minutes time to drive any route, X must be directly linked to some city in the core set, call it Y. The time to drive to X is then just the minimum time to drive to Y from City S (which you already know since Y is in the core set) plus the time to drive from Y to X. Now, you add X to the core set and record the time you have computed. If X is City T, then you are done. Figure 1 shows how the core set grows with each step of the algorithm.

> This was the first graph problem I ever posed myself and solved. The amazing thing was that I didn't publish it. It was not amazing at the time. At the time, algorithms were hardly considered a scientific topic.

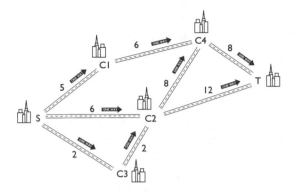

Figure 1
Dijkstra's shortest-path algorithm.
The original core set will consist of S alone.
Next, C3 will be added with a total of 2.
Next, C2 will be added with a total cost of 4 via the route S → C3 → C2.
Next, C1 will be added with a total cost of 5.
So the core set at this point consists of S, C1, C2, and C3.
Next, C4 will be added at a cost of 11 through the route S → C1 → C4.
At this point T will be added at a cost of 16 through the route S → C3 → C2 → T.

The mathematical culture of the day was very much identified with the continuum and infinity. Could a finite discrete problem be of any interest? Obviously the number of paths from here to there on a finite graph is finite. Each path is of finite length. You must search for the minimum of a finite set. Any finite set has a minimum—next problem, please. It was not considered mathematically respectable.

For many years, I had felt guilty about my lack of mathematical education. But eventually, I think I was glad I was spared the mathematical prejudices of the day.

The shortest-path algorithm, now known simply as Dijkstra's algorithm, has since been used in road building, routing through communications networks, and airline flight planning—any application in which one must find the best way to travel to a destination. Dijkstra soon twisted the algorithm slightly to solve a related practical problem.

Machine designers Loopstra and Scholten, who had been the engineers of the ARMAC, were building their next machine and sought

a way to convey electricity to all essential circuits, while using as little expensive copper wire as possible. Dijkstra solved the problem
with a method that he called, for technical reasons, the shortest sub-
spanning tree algorithm.

> Then I had two nice graph algorithms. Still there was no popular
> journal to publish it. I published it eventually in *Numerische
> Mathematik* in the first issue. Of course it had nothing to do with
> numerical mathematics.

It was an unusual paper for the time. For one thing, it proposed
an efficient way to solve a finite problem. For another, it carefully
proved its results.

> Almost all mathematicians at that time were in the teaching busi-
> ness. There was hardly any industrial mathematics. Then it was
> okay to leave something for the intelligent reader. At least that was
> what mathematicians felt; the standard way proofs were published
> was to publish a sketch of a proof.
>   It's quite clear that during those years of working on the early
> machines, I developed a set of quality standards, a set of values that
> were very different form the standard mathematical culture.
>   An emphasis on simplicity, completeness, correctness. It started
> with the writing of the Appalling Prose; a reference manual should
> contain everything; it should be complete and unambiguous.
> Machines are very unforgiving. They execute the program as is.
> The freedom of meaning one thing and saying something different
> is not permitted.

The Critical Section Problem

Dijkstra's innovative work on mutual exclusion and cooperating sequential processes began in the early 1960s with the designs of the
ARMAC's successors, the X1 and later the X8. From a hardware
point of view, they were typical machines of the day.

> The X8 had a big core storage cycle of 10 microseconds [about 100
> times slower than today's RAMs]. It had a red button and you

pushed it and the machine stopped. And if you pushed the green button, it would start again.

The software, by contrast, started a trend. Dijkstra arranged for each device attached to the computer to perform its tasks one step at a time, while exchanging messages with the computer. In computer jargon, these are called communicating sequential processes. Dijkstra started to think about ways to coordinate or synchronize these processes "so that I could reason about them."

Programmers face this challenge very often. Suppose two processes access the same data at the same time. One process might modify the data and therefore cause the other process to behave incorrectly. Avoiding bad behavior requires that while one of the processes accesses the shared data, the other one does not. This is what Dijkstra meant by synchronization.

Dijkstra thought again about trains—this time a train signaling system known as a semaphore. Suppose that there are two separate tracks between cities X and Y, one from X to Y and the other from Y to X. If the two tracks narrow to one during a portion of the route, then trains going in both directions must use the same piece of track. To avoid collisions, engineers use semaphores to ensure that only one train is on the shared track at any one time. The semaphores ensure that there is a green light in only one direction at a time and that the lights don't change color while there is a train on that critical piece of track. Thus use of semaphores ensures that only one train will be on the critical track at a given time. This is called *mutual exclusion*.

Dijkstra applied the notion of mutual exclusion to the communication between the computer and its attached keyboard. These two devices exchange information through a communication area in memory known as a buffer. The basic rule is that only one of these two should be reading or writing the buffer at a time.

> I realized that the coupling between the typewriter [the keyboard with its circuitry] and the machine was totally symmetric. Just as the machine would be forced to wait while the buffer was still full, so the typewriter would be forced to wait while the buffer was still empty. I knew we had a logical symmetry. I remember it was very liberating, very refreshing.

The cooperation of a number of units each with its own speed and clock—that was the given of the technology. What I wanted to do was arrange the cooperation in such a way that it would be independent of the relative speed ratios. I wanted to do this for safety's sake.

In 1961, Dijkstra thought of a way to represent the necessary protocols using two operations suggested by the railway semaphore: P and V. P stands for *passeren,* which in Dutch means "to pass," while V stands for *vrijgeven,* meaning "to give free." It is a testimony to the power of this idea that computer designers still use these letters, despite the dominance of English in computer science. Dijkstra's elegant solution came from his desire for clean reasoning.

The invention was not so much the P and V operations. The greater jump consisted of the decision to ignore relative speeds and to make reasoning about a system independent of that. That is not something that is taken for granted. What I remember is resistance to that idea. People found it difficult to accept that knowledge about relative speeds should be ignored.

That surely is no longer true. Virtually all modern processors and most memory boards support the functionality of P and V in hardware by means of a *test and set* instruction or something similar. These instructions either lock a computer resource, such as a buffer, and return success or determine that the resource is already locked and therefore return failure. IBM's 360 architecture was one of the first to implement test and set in 1964, thus giving the entire idea legitimacy.

Dining Philosophers

Dijkstra had been able to see things differently from his peers as a result of what he calls a "happy fact of my isolation." His ability to pose fundamental problems that others overlooked became apparent once again in 1965.

In the fall of that year, Dijkstra sat down one evening and prepared a now-famous examination problem for his students at the Einhoven Technical University. Dijkstra called it the dining quintuple problem, but it soon became known by the name Oxford Professor C. A. R. Hoare gave it: the dining philosophers problem.

Imagine that five Hunan philosophers are sitting around a table. Each one has a bowl full of rice, and a chopstick on either side of the bowl. The right chopstick of each philosopher is the left chopstick of his neighbor. (See Figure 2.) Now, the rules for dining are as follows:

1. Each philosopher thinks for a while, eats for a while, and then waits for a while.
2. To eat, a philosopher must hold both his right and left chopsticks.
3. The philosophers communicate only by the lifting and lowering of chopsticks. (They can neither speak nor write.)

Suppose that each philosopher uses the following algorithm in order to eat:

(i) Pick up right chopstick when available (waiting if right neighbor has it).

(ii) Pick up left chopstick when available (waiting if left neighbor has it).

(iii) Eat.

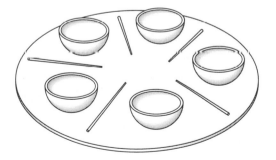

Figure 2
Dining philosophers problem. Each philosopher has one plate and one chopstick to his left and one chopstick to his right. To eat, he must hold both his left and right chopsticks, thereby preventing both of his neighbors from eating. The problem is to figure out a method by which each philosopher can eventually eat.

Several things can happen. If all philosophers decide to start eating at the same time, then they will all succeed at step (i), but will wait forever at step (ii). This situation is called "deadlock."

Seeing all his fellow philosophers holding only one chopstick, a philosopher waiting at step (ii) might put down his right chopstick and sit quietly for a while, watching his right neighbor eat. This gives rise to the possibility that an altruistic philosopher might never eat. This situation is called "starvation."

Even if all the philosophers do eat, some may eat more often than others. This situation is called "lack of fairness." Or life.

Variants of the dining philosophers problem crop up frequently in computer networks. For example, computers on a local area network often share a wire or broadcast channel over which only one message can be sent at a time. If all sites try to send at the same time, they all fail. If they then try again right away, they fail again. This is similar to deadlock. If one site always gets preference, then another site may starve or the protocol will be unfair.

For both philosophers and networks, one solution is randomization, as Michael Rabin will show us in the next chapter. If a philosopher or site cannot obtain a resource it needs, it waits an amount of time determined by some random process (e.g., a scintillation counter), then tries again. This scheme may still lead to starvation, since the process may be forever randomly unlucky, but the probability of such an event is small.

Several years after Dijkstra posed the dining philosophers problem, he was surprised to find that the designers of one of the most sophisticated computer systems then extant, M.I.T.'s MULTIX, had not thought about deadlock at all, and the system would, on occasion, abruptly stop—like so many philosophers each holding a single chopstick. With gentle irony, Dijkstra muses:

> You can hardly blame M.I.T. for not taking notice of an obscure computer scientist in a small town in the Netherlands.

Dijkstra in the United States

When Dijkstra accepted a job as Research Fellow for Burroughs Corporation, he used the position to push for verifiability in programming. But the discipline he advocates has proved to be distinctly unpopular, sometimes for cultural and sometimes for economic reasons.

> I think it was in 1970 that I gave my first talk in a foreign country on the design of programs that you could actually control and prove were correct. I gave the talk in Paris and it was a great success. On the way home, I gave the talk to a company in Brussels. The talk fell completely on its face. It turned out that management didn't like the idea at all. The company profits from maintenance contracts. The programmers didn't like the idea at all because it deprived them of the intellectual excitement of not quite understanding what they were doing. They liked the challenge of chasing the bugs.

Dijkstra's prodding, combined with the demand for high-quality software, has made the software industry significantly more disciplined. The one-line Dijkstra sound bite that every programmer knows is "GO TO considered harmful." GO TO causes a program to go from working on one job to working on something completely different without any plan for returning for the first job. Programs with many GO TO commands tend to be about as easy to follow as a legal contract in a Marx brothers' film.

Nevertheless, for many programmers, free-thinking hacking in a white heat of inspiration remains the ideal. Dijkstra views that kind of approach as a pathology.[1]

> People get attached to their sources of misery—that's what stabilizes many marriages.

Deep in the Heart of Mathematics

In the early 1980s, Dijkstra and his family moved to Austin, Texas, where he was awarded the Schlumberger Centennial Chair in Computer Sciences at the University of Texas. Now that their children are grown, the couple enjoys traveling in a Volkswagen camper they've

---

[1] Such habits impose an economic cost in a world in which roughly 40 percent of all software projects are canceled. About 70 percent of the rest are late.

nicknamed the Touring Machine. In his intellectual travels, Dijkstra has returned to mathematics in his quest for rigor.

> I am working on streamlining the mathematical argument: making the argument simpler, cleaner. It's really trying to transfer experience from programming into the wider area of mathematics.
>
> We all know that if you want to make something big, you have to compose it out of components—modules of some sort. You must be able to isolate parts. . . . It's well known from programming that this is not just a matter of division of labor because if you choose the wrong interface or an inappropriate one, the work explodes by a factor of ten—it's not just a sum.
>
> As an example, I have four composers living in different towns and they decide to compose a string quartet. You do the first movement, you do the adagio, you do the finale. Another way of dividing it is you do the first violin, you do the cello, you do the viola. In that latter distribution, an enormous amount of communication would be necessary between composers. That's a nice example of a practical and an impractical division of labor. Programmers have to think about this. A well-engineered mathematical theory has all the characteristics of the practical division of labor. The standard reaction of the inexperienced theorist who has demonstrated a complicated argument is to fall in love with that argument.

Dijkstra is impatient with definitions that cause problems for the reader. He once stopped reading Winston Churchill's *A History of the English Speaking Peoples* because "It was unnecessarily complicated. He would refer to the same people under different names." To Dijkstra, good definitions and a well-crafted argument are as essential as the idea itself.

> Whenever you are developing something new, you have tasks. You have to create a new subject matter. You have to create a language which is appropriate to discuss the subject matter. Many people are insufficiently aware of that second obligation.

Dijkstra's latest crusade for formalism in computer science and mathematics has spawned a book, *Formal Development of Programs and Proofs*. His central thesis equates writing programs with writing a clear proof. Research on this thesis and related topics takes place in summer meetings, one near Munich and one in Glasgow. Americans are, for the most part, conspicuously absent.

> Americans have a pathological fear of formal manipulation. It seems that the United States has a century of demathematicization which of course is very tragic because in that same century the mathematical computer is invented which is a major mathematical challenge. Somehow or other the mathematical nature of the challenge seems to have been ignored here as politically unpalatable.

Often Dijkstra appears as a Nabokovian character, a cultured European in a land of cowboys. He is for that reason a controversial figure—younger computer scientists, while awed by his accomplishments, wonder if he is concerned with things that still matter. By the same token, Dijkstra has doubts about the topics chosen by his colleagues. For example, when asked where artificial intelligence fits within computer science, he replies wryly, "Not." To some extent, he sees artificial intelligence as a manifestation of American naiveté.

> The Europeans tend to maintain a greater distinction between man and machine and have lower expectations of both.

There are two sides of Edsger W. Dijkstra. On one side, we see the creative scientist whose love of good problems and enduring solutions have made enormous contribution to the science and practice of computing. On the other side, there is the impatient observer of human folly whose acerbic pen (often a Montblanc) has alienated many colleagues. Dijkstra insists that he is an easy person to understand.

> I'm very constant in my opinions and judgments—frighteningly so.

To young scientists seeking insight into his research methods, he offers three golden rules.

1. Never compete with colleagues.
2. Try the most difficult thing you can do.
3. Choose what is scientifically healthy and relevant. Don't compromise on scientific integrity.

Michael O. Rabin

THE POSSIBILITIES
OF CHANCE

*We should give up the attempt to derive results
and answers with complete certainty.*

—MICHAEL O. RABIN

igital computers do exactly what they are told—no more, no less. Programming languages often underscore this point by providing imperatives like "do," "assign," and "begin" without so much as a please or a thank you. The subtext is: the machine is your slave; tell it your bidding. Given such encouragement, few people would imagine allowing the computer to make guesses or, even worse, to behave in a manner determined by chance.

Michael Rabin has imagined even stranger things. In the process of theorizing about how the computer might behave, he founded entire subdisciplines of computer science. His work has challenged mathematicians and computer scientists to accept the notion that certain computer programs *should* be designed to produce errors on rare occasions, e.g., asserting that a number is prime when in fact it isn't. Surprisingly, the rest of the computing world has agreed with him and such programs run billions of times a day in applications ranging from cryptography to robotics to communications.

Michael Rabin was born in Breslau, Germany (Wroclav, Poland, since World War II), in 1931, the scion of a long line of rabbis. His father, a rabbi and Ph.D. originally from Russia, taught Jewish history and philosophy at Breslau's then famous Theological Seminary, where he was the rector (academic head). Rabin's mother, who began composing children's stories when she was 17, held a Ph.D. in literature. Anticipating trouble, the family left for Palestine in 1935.

> My father, with his Russian background, knew what anti-Semitism was and that it might become mortally dangerous.
>
> He was always a Zionist. When Hitler came to power, he realized there was no future. He went to the board of the theological seminary and proposed to move the Seminary to Jerusalem. But those German patriots said that these were difficult times for the Fatherland, and they were not going to do something unpatriotic.

Rabin's father became a high school principal in Haifa soon after the family arrived in Israel. Michael began a religious elementary school.

> My sister, who is five years older than I, brought home *The Microbe Hunters* by Paul de Kruif. Reading that and other books on the pioneers of microbiology sparked my imagination, so that from the time I was eight until about twelve, I thought I might become a microbiologist.
>
> Then one day serendipity played a role and I was kicked out of class. There were two ninth-grade students sitting in the corridor solving Euclid style geometry problems. I looked at what they were doing. There was a problem they couldn't solve. So they challenged me and I solved it. The beauty of that, the fact that by pure thought you can establish a truth about lines and circles by the process of proof, struck me and captivated me completely.

Rabin tried to persuade his father to send him to the Reali School, which was modeled after the German Gymnasium but specialized in the sciences. His father wanted him to go to a religious high school, but the son prevailed.

> He correctly predicted that fascination with the exact sciences was going to drive me away from religion. I promised him it wouldn't be the case, but actually that was what happened.

Rabin graduated from the Reali School and entered the Hebrew University in the early 1950s just as the first articles about computers were being published in Israel.

> I got a book by S. C. Kleene called *Metamathematics*. It had a chapter on Alan Turing's work—mainly the notion of computability and the Turing machine. That really appealed to me a lot.
>
> Alan Turing gave an *a priori* definition of what is computable. Thus computing was not just a new technology; it was a technology that had a mathematical, logical idea as a foundation.
>
> Even at that time, I knew that I was going to be interested in logic, actually computability. But in order to vary my experience, I did a master's thesis in algebra, solving a problem due to Emmy Noether, in the theory of commutative rings.

Rabin would build on Turing's work later.

Turing and Computability

The English mathematician Alan Turing (1912–1954) defined the logical foundations of computation in 1935 before computers even existed. He used the word "computer," but in Turing's time the term meant a man or woman (they were often women, in fact) who was hired by companies to perform calculations.

Turing imagined how such a computer might perform a task. The computer would have a pencil and an infinitely long piece of paper divided by vertical lines into frames. Turing called the paper a "tape" and called each frame a "square." He wrote: "The behavior of the computer at any moment is determined by the symbol which he is observing, and his 'state of mind' at the moment."[1] Based on this observation and state of mind, the computer does one of three things: move to a neighboring square, change the current square, or enter a new state of mind. (We'll give some examples of states a bit later.) Turing assumed that the computer would use a finite alphabet, because he would not be able to distinguish among an infinite num-

[1]"On computable numbers with an application to the *Entscheidungsproblem*," *Proceedings of the London Mathematical Society*, (2)42, 1936–1937: 230–267.

ber of letters. Turing also assumed that the computer had a finite number of possible states of mind.

To Turing, a computer program consisted of a precise set of instructions given to such a person relating actions to observations. For example, "If you read 0 in the square you are scanning, replace it by 1 and move one square to the right." He hypothesized that the notion of what is computable is equivalent to what his imaginary computer could do with this long scroll of paper. As far as any computer so far built is concerned, he was right.

Turing hoped to resolve what the German mathematician David Hilbert in 1920 had called the *Entscheidungsproblem* (literally, "the decision problem"): Hilbert wanted a systematic means to prove or disprove any statement in mathematics that used the mathematical language known as first order logic. Turing used his machine to show that no such systematic means could exist. That is, he showed that there were problems that could not be computed.

For inspiration, Turing turned to the results of the Austrian mathematician Kurt Gödel, who in 1931, at the age of 25, had shown that there have to exist statements about arithmetic and euclidean geometry that can neither be proved nor disproved.

Gödel's idea was to take numbers and place them in logical sentences such as a sentence S which says: "S is not provable." He then reasoned as follows: Suppose sentence S were provable. Then S would be false, contradicting the fact that one had a proof of its truth. So, S is not provable; therefore it's a sentence that is true but not provable. (The idea of using a self-referential sentence dates back to the liar's paradox of Epimenides of Crete. It went like this: "Cretans always lie." So, was Epimenides a liar?)

Turing applied this paradox to programs given to his human machine. Turing first showed that there must be an infinite number of problems that are not computable. That is, there are an infinite number of problems that no computer can ever be programmed to solve. Technically, he showed that the number of mathematical functions was uncountably infinite whereas the number of programs was only countably infinite. What does all this mean? There must be an infinite number of functions out there without corresponding programs.

Turing went on to exhibit a particular problem, which he called the "halting problem," that could not be solved by a computer. The halting problem asks if Computer X can be programmed to decide in finite time whether Computer Y, loaded with another program and some starting data, will ever stop. Using reasoning akin to the liar's paradox, Turing showed that it is impossible to write a program for Computer X for this general problem. As Rabin points out, this has very practical implications.

> Suppose that a manager at IBM would come to somebody and say to her: Mary, the company needs an automatic way of checking whether programs stop or not, and I want you to find an algorithm for it—to find a computational method for answering this question. Well, Turing showed that it is not possible.

At the time Rabin became interested in these issues in the 1950s, Israel had no computers and few people even working on computation. So Rabin moved to the United States, first to the University of Pennsylvania to study mathematics, then to Princeton for a Ph.D. in logic.

Rabin became a student of Alonzo Church: "He was a very rigorous man who enabled me to really understand aspects of mathematical logic." Church was also the proponent of a theory of computability different from Turing's but equally powerful.

Rabin's Ph.D. thesis addressed questions of the computability of algebraic groups. *Groups* are fundamental mathematical structures with applications in many sciences, especially theoretical physics. Rabin showed that many of the problems concerning groups could not be solved by a computer.

> Turing defined the great divide between what is computationally solvable and what is not. My thesis showed in a similar vein that certain problems pertaining to groups were not computable.
>
> The group is given in a certain way and you want to ascertain whether it has certain properties; for example whether it is commutative or not. [An operation, denoted, say, by *, is commutative if it is true for any elements A and B in the group that A * B = B * A.]
>
> In that question, the group itself is described so it looks like a program. That question is not computationally answerable. There were many other examples.

Computers That Guess

In 1957, while Rabin was writing up his thesis results, IBM Research offered summer jobs to him and another young logician named Dana Scott. The company left them free to do whatever struck their fancy, and the two collaborated on what was to become a fundamental theorem in computer science. As part of their work, they proposed the notion of a computer that could "guess" solutions.

Rabin and Scott began by considering a limited form of Turing's theoretical computer: one that was forbidden from writing on the tape. Such a computer, also known as a *finite state machine,* records what it learns in a memory whose number of states is fixed once and for all.

An everyday example of a finite state machine is a combination lock. If the lock requires three numbers in sequence, then we can imagine that its states correspond to no numbers dialed, the first number correctly dialed, the first two numbers corectly dialed, and all three numbers correctly dialed. Only the last state releases the lock. An incorrect turn along the way returns the state machine to the no-numbers-dialed state.

A finite state machine has many uses in computer applications. For example, it can determine which parts of a text match Br*e C*n, where the * can match any sequence of nonblank characters. It would find matches, for example, in Bryce Canyon as well as Bruce Chatwin.[2]

Rabin and Scott were intrigued by an implicit limitation of Turing's model: a machine with a given set of instructions and a

[2]But it has limitations, too, as Noam Chomsky of M.I.T. had already shown in 1957. Chomsky showed that finite state machines are too weak to model English grammar. He argued that "if-then" sentences in English can be arbitrarily nested; e.g., if sentence S1 is true, then sentence S2 is true also. Here, S1 and S2 can be arbitrary sentences having ifs and thens of their own. Therefore, he said, any grammar for English must recognize sequences containing elements "if then," "if if then then," "if if if then then then," and so on. He showed that finite state machines are unable to handle such sequences. This led him to introduce the context-free grammars that Backus used.

It also led Chomsky to argue that B. F. Skinner's condition-response model of learning was not powerful enough to teach English grammar, precisely because that model was equivalent in power to a finite state machine.

particular input will always behave in the same way. Its behavior is "deterministic."

In Turing's language, a deterministic human computer will pass through the same "states of mind" every time it is presented with the same sequence of inputs. Such determinism was a basic tenet of Alan Turing's work. To explain nondeterministic behavior, Rabin offers a dinnertime analogy.

> We postulated that when the machine is in a particular state and is reading a particular input, it will have a menu of possible new states into which it can enter. Now there is not a unique computation on a given input, because there are many paths.
>
> Consider a menu in a fancy French restaurant. There are a number of hors d'oeuvres, soups, and fish selections before the main course. Then a number of meat main courses and so on down the line. You are in the start state for selecting a meal. In the first instance, you select the hors d'oeuvres, and then you select the soup, and so on. This is not a random process. We are not tossing a coin. But it is nondeterministic because you have choices as to the next state (i.e., menu item selection) that you move into.
>
> The choices are the states. Each of these sequences of choices represents a possible computation and results in either satisfaction or dissatisfaction with the meal. Acceptance or rejection.

The two logicians recognized that choice can lead to ambiguity, so they said that a "nondeterministic" machine "accepts" a sequence if *at least one* of the possible computations from the input reaches an "accepting state." Using the terms of Rabin's analogy, if a menu selection provides a nondeterministic patron with a good choice for every course, then he will say that the restaurant is good. Otherwise, he will say that the restaurant is bad. A guessing or nondeterministic machine might seem to be a mere intellectual curiosity, except for two things.

First, Rabin and Scott showed that for finite state machines, any problem solved with a nondeterministic machine can also be solved with a deterministic one, though many more states may be needed. They showed how to convert a machine from one type to the other.

Second, nondeterministic finite state machines turned out to be an excellent way to express pattern searches in language translation, library document searching, and word processing programs. In fact, every time you do a pattern search with your computer, it probably uses some variant of Rabin and Scott's procedure to find the matching text in your file. Rabin and Scott worked all this out at Olympic speed.

FINITE STATE AUTOMATONS

The patterns processed by a finite state machine need not be English words; they can be combinations of letters or numbers or other symbols. Since computers work in a binary number system, we'll use an alphabet of just two symbols, "a" and "b."

A SIMPLE FINITE STATE AUTOMATON

The machine shown in Figure 1 accepts the input if it is "ab" but rejects any other input.

The machine begins in the start state and proceeds to state 2 if the first symbol is "a" or to state 3 if the first symbol is "b." From state 2, the machine proceeds to the final state if the second symbol is "b" or to state 3 if the second symbol is "a." (The double circle is the conventional symbol for the final state.) The machine will leave the final state if it receives further input. From state 3, there is no exit. So, this machine will accept "ab" and only that. If the automaton reads the last symbol of the input and ends in that state, then the input is accepted. Otherwise, it is rejected.

This machine is deterministic because from each state there is only one transition corresponding to "a" and only one corresponding to "b."

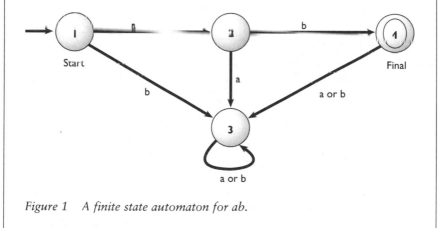

Figure 1   *A finite state automaton for ab.*

STRINGS OF ANY LENGTH

The simple illustration above was just a quick warm-up. A finite state machine can also accept strings that are of arbitrary length. Figure 2 is a finite state machine that can accept the pattern consisting of strings "ab," "abab," "ababab," "abababab," and so on. Any other string will be rejected.

As you can see, this machine is the same as the first one except that it has a transition from the final state to state 2 on an input of "a." The path from state 2 to state 4 and then back to state 2 allows the automaton to accept strings of any length. For example, on the pattern "abab," this machine will start in state 1, go to state 2 on the first symbol "a," the final state on the second symbol "b," back to state 2 on the third symbol "a," and back to the final state on the fourth symbol "b."

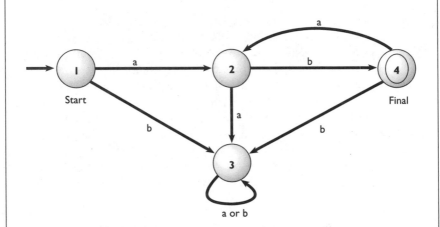

Figure 2   A finite state automaton for ab, abab, ababab.

STRINGS OF THE FORM
AB, ABAB . . . OR ABBA, ABBABA . . .

Now consider the problem of recognizing the pattern consisting of either "ab," "abab," "ababab," "abababab" . . . or "abba, "abbaba," "abbababa". . . . The difficulty is that both infinite patterns start with "ab," so the machine must choose a pattern to work on. This choice is easy to represent using a nondeterministic finite state machine.

In Figure 3, the nondeterministic choice occurs at state 1, the start state, where an input of "a" allows the machine to go either to state 2 or state 3. According to Rabin and Scott, if there is a path that leads from the start state to a final state such that the first transition along the path has the first letter of the input, the second transition has the second letter, and so on, then the input is accepted. For example, "ababab" is accepted: a path goes from the start state to state 2, then 4, then 2, then final state 4, then 2, and then final state 4 again.

As a second example, "abba" is accepted because of a path going from the start state to state 3, then 5, then 6, then final state 7. On the other hand, "abaa" will not be accepted since there is no path from the start state to a final state for that input.

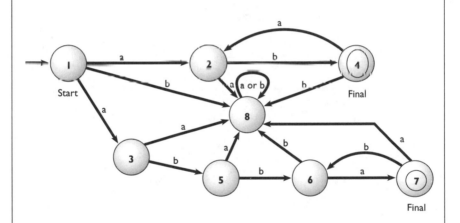

Figure 3   A nondeterministic finite state automaton that will accept strings of the form ab, abab, abababab, abababab, ... or of the form abba, abbaba, abbababa, ... but no others.

CONVERTING A NONDETERMINISTIC MACHINE TO A DETERMINISTIC ONE

The nondeterministic finite state machine is easy to write, but how should one program it? Rabin and Scott came up with a systematic procedure based on set theory. Figure 4 shows how to write a deterministic finite state machine that will accept strings of the form "ab," "abab," "ababab," "abababab" . . . or of the form "abba," "abbaba," abbababa," . . . but no others.

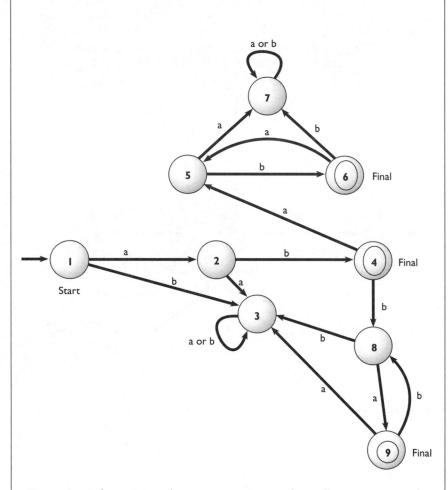

Figure 4   A deterministic finite state automaton that will accept strings of the form ab, abab, abababab, abababab, . . . or of the form abba, abbaba, abbababa, . . . but no others.

That was a fabulous collaboration. . . . One of us would formulate a question and then we would go to our respective corners and the other one would come up with a solution. Maybe overnight. Within maybe three weeks we had all the answers, including many results that did not go into the paper because we didn't want to make it too long. These were later rediscovered by other people.

Their paper did not appear in the technical literature until 1959. For the next ten years, theoreticians extended its findings in various ways. But the notion of nondeterminism proved to have even wider appeal. Rabin was surprised.

I must say we didn't see all the implications of this. Our notion of nondeterminism was for us a mathematical creation. I've done a lot of consulting in industry since then. I may sit listening to practical systems engineers, say operating system experts, discussing the design of an operating system and somebody says "I want you to specify whether this is deterministic or nondeterministic." I then smile to myself because this harks back to these mathematical abstractions of nondeterministic machines.

The Inherent Difficulty of Computation

In the summer of 1958, Rabin returned to IBM's Think Tank-on-the-Hudson, the Lamb Estate. John McCarthy was there struggling with list processing in Fortran and posed a puzzle to Rabin:

There are two countries in a state of war. One country is sending spies into the other country. The spies do their spying and then they come back. They are in danger of being shot by their own guards as they try to cross the border.

So you want to have a password mechanism. The assumption is that the spies are high caliber people and can keep a secret. But the border guards go to the local bars and chat—so whatever you tell them will be known to the enemy.

Can you devise an arrangement where the spy will be able to come safely through, but the enemy will not be able to introduce its own spies by using information entrusted to the guards?

Rabin came up with a solution using what is now known as a *one-way function*. One-way functions translate into computational procedures. They are easy to compute in one direction, but hard to compute in the other. Familiar functions are not like this. For example, tripling a number is a two-way function, because it is easy to calculate $3 \times y$ and also easy to calculate $z/3$. Rabin used a one-way function developed by the mathematician John von Neumann.

> Suppose you take a 100-digit number, x, and square it. This is easy to do by computer. You get a 200-digit number. Out of these 200 digits, you take the middle 100. Call the resulting number y. Now if I give you x, you can calculate y. But if I give you y and ask you to calculate x, you have presumably to try all the possible xs, of which there are many. So "middle squaring" is a function which is easy to compute and difficult to invert. The solution to the spy password problem was by means of a one-way function.

Do you see how? (If you like puzzles, stop here and think about it before reading on.)

Rabin's solution goes like this: suppose the guards know how to compute the middle square and suppose they have a list of y values, one for each spy. When a spy comes to the border he gives his x value and his name. The guards check that the middle square of x is y for that spy (and check that the spy has not entered anywhere else). A would-be imposter would never be able to find the appropriate x, even if a guard gave a list of all the y values away to some temptress at a bar.

The more general question arising from Rabin's solution was how to *prove* that a function such as middle squaring is difficult to invert. That is, no matter how one does it, the minimum number of operations is very large. This led Rabin to a general investigation of the minimum amount of work needed to perform a given computational task, i.e., the inherent difficulty of that task. In the physical world, as Rabin point out, this notion is quite intuitive.

> Suppose that I have a book on this table and I want to raise this book to the level of the ceiling. There are many ways of doing it. I could pick it up by hand, get up on the chair, and lift the book to the ceiling. I could create a pulley, attach it to the ceiling with a string, and pull the book up. I can use an electric motor and so on.

But there is a certain inherent minimum amount of work required for the task by any of these methods. Namely, the weight of the book times the height of the ceiling. Defining such inherent difficulty is what I did for computational tasks.

To illustrate how this idea has been applied, consider a game in which you are to determine via yes or no questions which number between 1 and 1000 someone else is thinking of. A well-known strategy is to ask whether the number is greater than 500. If so, then is it greater than 750? If not, then is it less than 250? And so on. Each question reduces the set of possibilities by half. This strategy is known as *binary search*.

The binary search strategy requires 10 questions before you know the answer to this game for sure. You might wonder whether any strategy would produce quicker results. Imagine that you choose some strategy that does not guarantee to reduce the set of possibilities by half each time. For example, you might first ask whether the number is less than 100. If your interlocutor says yes, you will be able to determine the correct number in fewer than 10 questions. On the other hand, you could be unlucky and your interlocutor might answer negatively, implying that the number is between 100 and 1000. Thus you can show that the best *guaranteed* outcome you can hope for is one that reduces the original set of possibilities by half with a single yes or no question. From this it follows that 10 is the minimum number of yes or no questions necessary to guarantee a correct answer. This, then, is the minimum work necessary.

> I showed that no matter how the measure [of computational difficulty] is defined or how the measure is being used, there are always computable functions that are very, very difficult to compute, functions which require a lot of work in terms of that measure. Thus, I showed it was a meaningful concept because functions or computations do differ in terms of their inherent difficulty.

With Michael Fischer of Yale University, Rabin used this theory to look at mathematical problems involving only addition called Presburger arithmetic. In the 1930s, the Polish mathematician M. Presburger had shown that determining the truth or falsity of statements in his system was computable with a Turing machine.

Presburger was not interested in the number of operations it would take to perform these computations, only in whether the correct answer could be found in finite time.

Rabin and Fischer were able to show that determining the truth or falsity of statements in Presburger arithmetic could be impractically hard—doubly exponential in fact. Statements of just 100 symbols could keep a trillion computers busy, each performing a trillion operations per second for a trillion years without coming close to solving the problem.

As it happened, writing programs to prove theorems in Presburger arithmetic was a goal in the artificial intelligence community just when Rabin was thinking about it.

> I was invited to give a lecture in Stockholm at the International Federation of Information Processing. I entitled the lecture "Theoretical Impediments to Artificial Intelligence."
>
> My lecture actually occurred on the day that Nixon abdicated [in 1974]. I thought there wouldn't be an audience. During the lecture before mine, the hall was almost empty. Okay, Nixon. Then people all of a sudden started streaming in—it was really a sight.
>
> So I described results like the intractability of Presburger arithmetic—some AI people were writing theorem-proving programs and here was a result, in a very simple case, proving that this is absolutely hopeless.
>
> The theorem-proving people were extremely worried. They were afraid—they told me so—of the impact this would have on the funding agencies.
>
> Then at the end of that lecture, people lined up in front of the microphones to direct questions to me. All they wanted was to elicit a statement that this wasn't the end of the world.

About the same time as Rabin's speech, Steve Cook, Leonid Levin, and Richard Karp independently identified a huge collection of less difficult but still seemingly intractable problems called NP-Complete problems (see the coming chapter on Cook and Levin). A sense of pessimism hung over the community: it was as if the field of computer science would be circumscribed by problems which were ostensibly simple but which were just too difficult to solve by deterministic methods.

Through his work on the inherent difficulty of problems, Rabin had helped to tie a computational Gordian knot. Next, he would provide an idea that would cut through part of it, making the solutions to some problems feasible.

Computers that Toss Coins

In the Stockholm lecture, I raised the question as to what can one do, given such results about inherent difficulty?

I proposed that we should give up the attempt to derive results and answers with complete certainty. We should use randomness in a certain way and get the results more quickly but with a small probability of error. At that time [1974], I had examples of this method, but they were somewhat contrived.

In 1975, I came on sabbatical to M.I.T. I came across a result by Gary Miller, who showed that using an unproven assumption, the so-called Riemann hypothesis, one can test very large numbers for primality by ordinary deterministic algorithms.

A *prime* is a whole number that can be divided only by itself and 1; for example, 2, 3, 5, and 7 are primes. The number 15, which can be divided by itself, by 1, by 3, and by 5, is a *composite* number. The Miller result was important to mathematicians because the previously best known methods to test for primality took inordinately large amounts of time. Such methods required testing a significant fraction of the numbers between 1 and the square root of x to determine whether x is a prime.

Rabin took Riemann's hypothesis and proved a probabilistic statement about it. The result was the fastest test to find prime numbers that there was—and is.

Underlying the primality test is the concept of a "witness" to the compositeness of a number. Consider the number 143. If I give you the number 13, you can divide 143 by 13, concluding that 143 is 11 times 13, so that 143 is composite. Thus 13 is a "witness" providing proof of the compositeness of 143; similarly for the other factor, 11. The trouble is that many numbers such as 143 have just a few witnesses. One is unlikely to find them by chance.

I have devised another type of witness to compositeness: numbers which constitute proof that a number n is composite if they stand in a certain relationship to n. For an n which is composite, we can prove that at least three-fourths of the numbers between 1 and n are witnesses in the new sense to the compositeness of n.

Rabin's new kind of witness used Miller's results as well as powerful ideas in number theory concerning the density of primes. The result was a remarkably simple algorithm.

The randomized test for primality of n now proceeds as follows. Randomly choose, say, 150 numbers between 1 and n. Check for each of these numbers whether it is a witness, in the new sense, to the compositeness of n. If any one of the chosen numbers is a witness, then we know that n is not a prime. The novel aspect is that if none of the chosen numbers is a witness, then we declare that n is a prime.

Can it happen that even though n is composite, we erroneously declare it to be prime? Yes. But for this to occur it must happen that, even though at most one-fourth of the numbers between 1 and n are nonwitnesses, in randomly drawing numbers in this range 150 times, we come up with a nonwitness every time. This event has a totally negligible probability [smaller than one in a trillion trillion trillion trillion trillion trillion].

Rabin decided the approach needed some experimentation, so he showed the algorithm to a colleague at M.I.T., Vaughan Pratt, who volunteered to program it.

So he programmed it and we tried it on some very large known primes and nonprimes, and it produced the correct results. Then we tested it on numbers which hitherto were inaccessible to any effort.

Vaughan was a very hard worker and was at his computer terminal late at night some time in the winter of 1975. We had a Chanukah party at home with many guests and the usual things—latkes and so on. Close to midnight I get a telephone call.

Michael, this is Vaughan. I'm getting the output from these experiments. Take a pencil and paper and write this down. And so he had that $2^{400} - 593$ is prime. Denote the product of all primes p smaller than 300 by k. The numbers $k \times 338 + 821$ and $k \times 338$

+ 823 are twin primes [primes which differ by 2]. These constituted the largest twin primes known at the time. My hair stood on end. It was incredible. It was just incredible.

Perhaps only mathematicians can get the shivers over extremely large primes, but nonmathematicians can perhaps appreciate that prime numbers have a kind of universal quality to them. They are so natural that their discovery is lost in antiquity. Today they play a central role in the messages we encode on the Voyager spacecraft that have left the solar system on their way to alien civilizations. Why should primes be there? Why should there be relatively fewer of them as numbers get larger? Why doesn't there seem to be an easy pattern in their distribution? There are many more questions, and every mathematician yearns to make a fundamental discovery concerning primes. Pratt and Rabin had done just that. But could randomization do more?

In the culture of computer science, an idea that works in one situation is called a hack, an idea that works twice is called a trick, and an idea that works often and pervasively is called a technique. Where would randomization fit? Shortly after his discovery, Rabin gave a lecture about his findings at Carnegie Mellon University.

> After the lecture, I stood in the hall and there were people standing around me in a semicircle and saying that it was very nice, but the consensus of opinion, with one exception, was that this was very specialized, that I was using specialized properties of prime numbers, namely the theorems about witnesses and specialized properties of the geometric problems that I was studying to yield the solution of these two particular problems. This was not, people said to me, of wide utility.
>
> The only dissenting opinion came from Joe Traub [then chairman of Carnegie's Computer Science department and a computational theorist]. He said that using randomization and allowing the possibility of errors is a new departure.

Traub was right. Randomization has had many applications in computational geometry (the field of algorithms most closely associated with robotics and manufacturing), distributed computing, information retrieval, cryptography, communication, and even computer hacking. Robert Tappan Morris, a student of Rabin's when an undergraduate at Harvard, used randomization to help his virus propagate through the Internet in November 1988.

In a scientifically benign application, special purpose networks connecting massively parallel computers use a randomized algorithm discovered by Leslie Valiant, one of Rabin's colleagues at Harvard. The straightforward way to use a network is to send a message directly from its source to its destination. Valiant was able to show that a way to reduce contention (traffic jams) is, instead, to have each message go from its source to a random site and then from that site to its destination. That such a seemingly crazy scheme should help is testimony to the power of randomness.

The most exciting—and controversial—application of randomization is to cryptography, specifically to a form of cryptography known as *public-key cryptography*. Classical cryptography uses private-key methods: the sender and receiver know the same key; the sender encodes the message with that key and the receiver decodes the message with the same key. The problem with private-key systems is that the sender must somehow send the key to the receiver beforehand. This requires a courier who might be captured, bribed, blackmailed, or lost.

Public-key cryptography is based on one-way "trap-door" functions, similar in effect to the function Rabin used earlier to solve McCarthy's spy puzzle. In public-key cryptography, a person X can broadcast key K1, permitting anyone to communicate with X by sending a message encoded with K1. Only X holds a secret key K2 that can decode such a message. Because K1 is created using a one-way function based on large prime numbers, nobody besides X can figure out K2 given K1 (we think). This eliminates the need for couriers and allows a much more spontaneous style of communication than private-key cryptosystems.

The U.S. National Security Agency (NSA), whose job it is to protect government secrets and to break other countries' codes, has certified only one encryption scheme: a method for constructing private-key codes that the NSA codeveloped with IBM, called the Data Encryption Standard. The NSA involvement has led critics to suspect that the NSA has the ability to crack codes using the standard.

In 1993, the NSA promoted a new encryption chip called the Clipper, designed with a "back door." Every chip will have a code key held in a government repository. If the government wanted to read your mail, they could petition for a subpoena to let them retrieve your key. The government's rationale is that they have the right, in cases of national security and law enforcement, to read private messages. (Matthew Blaze, a researcher at AT&T Bell Laboratories, showed in mid-1994 that the Clipper technology could be used to encode messages that law enforcement authorities would not in fact be able to crack. By the time you read this, Blaze's observation may have led to the Clipper-based system's replacement by another government proposal.)

Critics prefer the public-key encryption method called the Rivest, Shamir, and Adleman (RSA) algorithm. That method uses Rabin-style randomization.

First of all, the primality test itself plays a role in these public key cryptography systems. You start by producing very large prime numbers. Numbers which are products of two such primes constitute a public key.

The RSA algorithm produces two large primes P and Q and then creates a product P times Q, call it N. N and some other number is the so-called public key. The RSA method can be broken if somebody finds a quick algorithm for factoring very large numbers (i.e., finding the P and Q given N). But we don't know whether such an algorithm exists. So, the viability of those methods depends on the unproven intractability of factorization.

Will factorization be broken (or has it been)? Rabin smiles, but then says he prefers not to commit himself.

Random Pursuits

Rabin is currently the T. J. Watson Sr. Professor of Computer Science at Harvard and the Albert Einstein Professor of Mathematics and Computer Science at the Hebrew University in Jerusalem. He divides his year between the two universities, while his family remains in Jerusalem.

His wife is head of the international division of the Israeli Department of Justice. One of his daughters is a lawyer; the other daughter is a computer scientist whose work applies randomization to distributed systems and cryptography.

Rabin's own recent work applies randomization to ensuring reliability (with high probability) in large parallel computers. In this, he works in a team, mostly with Indian computer scientists Krishna Palem and Partha Dasgupta and two Israelis, Yonatan Aumann and Zvi Kedem. During his career, Rabin has shown that the computer can solve so many problems in so many varied domains that one wonders whether he sees any limits.

> When we talk about complex tasks, I think that we are completely lacking in understanding as of now. For example, we lack understanding as to how human memory works. If I say Beethoven to you, you immediately know that I am referring to a composer. One could construct a memory organization method, and a computer program, which would classify names in a limited field.
>
> But our memory is much more complex. You can walk in the street and bump into someone who is somewhat unwashed, and all of a sudden you remember someone who was sitting next to you in high school bothering you because he didn't wash properly.
>
> Even that is a simple example. We remember things by structure. You can see a person in a state of stress and that may remind you of a joke that is of an entirely different nature. We make these leaps all the time.
>
> We see a person from behind—someone walking in a certain way. And say that so and so was Jerry. We almost never make a mistake. At least I almost never make a mistake. So we need very, very little. How that is done we simply don't understand.
>
> I don't think this has to do with a difference between the power of the mind and the power of the computer. It is simply that we don't know how to write a computer program to do it.

Donald E. Knuth

BOUNDLESS INTERESTS,
A COMMON THREAD

Computer programming is an art form, like the creation of poetry or music.

—DONALD E. KNUTH

an Donald Knuth really be just one person? Included among the 150 papers he has written are three of the most important algorithms in the field. His magnum opus (he is currently writing its fourth volume), *The Art of Computer Programming,* includes original research and a survey of most of the field. Earlier volumes have spawned Chinese, Japanese, Russian, and Hungarian editions. Over the span of a thirty-year career, Knuth has found time to create powerful software systems for typography, to write on such diverse topics as ancient Babylonian algorithms and Biblical psalms, and to pen a novel. In his "spare time," he plays a pipe organ that he designed.

Throughout his career, Knuth has received public acclaim and awards, including computer science's highest prize, the Turing Award, in 1974, and the National Medal of Science from President Jimmy Carter in 1979. Yet Knuth regards the accolades with a certain detachment. The bowl which commemorates his Turing Award now holds fruit.

From Alfred E. Newman to von Neumann

Knuth was born in Milwaukee in 1938. His father, the first college graduate in the Knuth family, started as a grade school teacher, and later taught bookkeeping in a Lutheran high school. He also played the church organ on Sundays. Donald inherited his father's appreciation of music and education, particularly patterns in language.

> I was mostly interested in what the teachers were best at. We had very good training in the diagramming of sentences. A bunch of us would have fun after class figuring out the parts of speech in sentences of poetry.

As editor of the school newspaper, Knuth invented crossword puzzles. He remembers enjoying the search for patterns in words. Knuth began winning awards early on. When he was in eighth grade, a candy manufacturer sponsored a contest to see who could compose the most words out of the letters in the phrase "Ziegler's Giant Bar." Knuth decided to give it a try.

> I found approximately 4500 words without using the apostrophe. With the apostrophe, I could have found many more. The judges had only about 2500 on their master list.

He won first prize—a television set (a pricey item in those days)—and enough Ziegler candy bars for the entire school. In high school, Knuth won honorable mention in the Westinghouse Science Talent Search with an unusual proposal: "The potrzebie system of weights and measures." With the care that would mark his later career, Knuth defined his basic units precisely: the potrzebie, the thickness of *MAD Magazine* #26; the MAD, 48 things; and the whatmeworry, the basic unit of power. In June of 1957, *Mad Magazine* itself bought the piece for $25, the first publication of Donald's prolific career. But music, not writing or science, took most of his time during high school.

> I thought when I went to college I would be a music major. I played saxophone, but then the tuba player got into an accident and I became a tuba player. I arranged a piece for band that combined all kinds of themes off TV shows—Dragnet, Howdy Doody Time, and Brylcreem. I knew nothing about copyright law.

His plans to become a musician changed when Case Institute (later Case Western Reserve) offered him a physics scholarship.

The system channeled anybody with an aptitude for science into physics. It was post-World War II and there was a lot of excitement in the field.

In high school, Knuth found mathematics uninspiring. But at Case, Paul Guenther, who taught freshman calculus, persuaded him to switch from physics to math. Guenther became Knuth's mentor in the process.

I had never met a mathematician before. He had a good sense of humor, but no matter what you said to him, he was unimpressed.

In 1956, Knuth had his first encounter with a computer, an IBM 650—a pre Fortran machine. He stayed up all night reading the manual and taught himself basic programming.

The manuals we got from IBM would show examples of programs and I knew I could do a heck of a lot better than that. So I thought I might have some talent. I didn't realize that almost anybody could improve on those programs—the existing books were atrocious. When I was starting to learn computers, that's exactly when Backus was busiest with Fortran.

Knuth's first program factored numbers into primes. Another program taught the computer how to play tic-tac-toe. But that was just hacking around. In 1958 he wrote a program for the Case basketball team that rated each player based on such criteria as shots missed, steals, and turnover. The coach was impressed by the program and later claimed that it helped the team win a league championship. *Newsweek* wrote an article about the program and IBM used a picture of Knuth posing next to the 650 in publicity photos.

Knuth was fascinated by what he could make the machine do. It turned out the computer fit even his musical interests.

Mathematics is the science of patterns. Music is patterns. Computer science does a lot with abstract things and making patterns. Computer science, I think, differs from other fields most in that it constantly jumps levels—from looking at something in the small to looking at it in the large.

A lot of careers are based on a perceived need and people find ways to solve critical problems—medical careers, for example—while other careers like computer science are chosen because of the kind of mind structure that you develop as a child.

If you happen to be in a certain 2 percent of the population, you have these mental qualities that resonate well with computers, and you naturally gravitate to computer science. It's just a way of thinking that distinguishes us. Eventually, I learned that I was a computer scientist.

Knuth graduated summa cum laude from Case in 1960. By an unprecedented vote of the faculty, he also received a master's degree in mathematics at the same time. He continued at the California Institute of Technology where he completed his Ph.D. in math three years later. He wrote a thesis in combinatorial geometry: "Finite Semifields and Projective Planes."

After graduation in 1963, Knuth joined the Caltech faculty as assistant professor of mathematics, but he pursued his interest in computers. Since 1960 Knuth had consulted for the Burroughs Corporation. Burroughs (now merged into Unisys) was a leader in the computer industry and had established contacts with such luminaries as Edsger Dijkstra.

Knuth's work for Burroughs involved both hardware and software design, especially support for the newly created Algol 60 programming language. This work gave him a chance to know Dijkstra personally, thanks to their common interest in compilation. Dijkstra and J. A. Zonneveld had completed the first implementation of an Algol 60 compiler in August 1960.

> We met and wrote letters back and forth. His great strength is his uncompromising aesthetic. Me, I've always been wishy-washy. If he tells me he likes what I do, then he really likes it. If he tells me he doesn't like it, then he really doesn't like it. So that makes him a very valuable correspondent.
>
> There was a great gap between mathematics and computer science in those days. When writing a program, you twiddled until

you thought it would work. The idea of using mathematics to prove that a program would work—that was a radical notion, nobody conceived of it as possible. Dijkstra was one of the great pioneers of proving things about computer programs.

The Art of Computer Programming

In January of 1962, while Knuth was still a graduate student, the textbook publisher Addison-Wesley asked him to write a book about compilers, a barely understood topic of study at the time. He began that project the following summer, and tested the first drafts on his students at Caltech in the fall of 1963.

> By 1966 I had 3000 pages of draft written out and I started to type it. I compared the size of my handwriting to characters on a printed page and estimated that my 3000 pages would come down to 700. But the publisher said no, the ratio was one to one! After frantic meetings, we decided to plan for a series of seven volumes.
>
> A single person could surround the field of computer science pretty well in 1966. But it has grown and grown. I've done my best to try to keep up with it. Now I know that Volume 4 [about combinatorial algorithms] itself is going to be about 2000 pages long— Volumes 4A, 4B, and 4C—and I figure I can finish it in the year 2003.

For a freshly minted Ph.D. to write such a comprehensive text was surprising enough, but its reception was even more so: The first three volumes of *The Art of Computer Programming* became the textbooks of choice through the early 70s and are still frequently used as reference works. Their continued popularity results from Knuth's thoroughness in treating his subject. When probability theory is required, the book goes through all the details. After explaining an algorithm, Knuth presents a program that implements it—just to make sure there is no misunderstanding. Through it all he mixes rigor with wit and tries to show the beauty underlying every idea. As New York University compiler designer Ed Schonberg puts it, "Dijkstra taught us to tell right from wrong; Knuth taught us how to tell so-so from terrific."

Compilers

While writing his textbook, Knuth began his research into compilers. *A compiler* is a program that translates from one language (the source) to another language (the target). The source language can be a high-level language—any of the modern computer languages such as Fortran, Cobol, C, or even a word processing, graphical, or spreadsheet language. The target language is the sequence of 0s and 1s that a particular computer understands. In the early sixties, it was not at all clear how to do this translation. Backus's Fortran group at IBM wrote the earliest large-scale compiler in a four-year herculean effort. The subject of compilation was considered so difficult that compiler writing often had the highest course identification number of any graduate course.

> I got into compilers because I thought the most amazing thing you could do with computers was to have them write their own programs. When computing is applied to computing, that's when computer science reaches an ultimate completeness.
>
> Programmers of the 1950s would punch cards in algebraic notation and feed them to a machine. Then the lights would flash and punch, punch, punch—the machine was punching out computer instructions! It was amazing. I couldn't believe this was happening and I had to know how it worked. And when I found out, I saw that even better ways were possible.

In fact, modern software tools have made compiler writing much easier, so most schools today offer the same course to undergraduates. Knuth played a big role in the development of those tools.

> In compilers, my work that is most well known is LR(k) parsing, which occurred to me immediately after I finished the first draft of chapter 10. (Remember, I thought I was writing one book.) The fundamental idea came naturally, because I had just surveyed what was known before.

What Knuth discovered was a general method for the task of parsing—finding the structure for the translation. The parser's job is to take a string (a sequence of words) and determine which grammatical rules apply to the string so that it can be translated. For efficiency reasons, practical parsers perform this translation in one pass. That is, they never reverse decisions they have already made.

But sometimes it is difficult to make a decision. For example, consider the two English sentences "Flying planes is fun" and "Flying planes can crash." In the first case, the proper interpretation of "Flying planes" is "the act of flying planes"; in the second case, the proper interpretation is "planes that are in the air." These interpretations follow from the grammatical categories assigned to "flying" and "planes."

To solve the problem of parsing sentences like these, Knuth made use of an existing technique known as *lookahead*. Using lookahead, the parser, once presented with "Flying planes," will look farther ahead in the string to decide which grammatical interpretation to apply. Knuth's algorithm could handle many more languages than previous methods.

Attribute Grammars

Following his work on parsing a programming language statement, Knuth went on to work on a general method for finding the meaning of programs. He was looking for a formulation as elegant as what Backus and Naur had found for syntax.

> Backus's work was of great interest to me. Backus realized that computer languages had this nice structure and that an elegant, formal syntax was possible. I was quite fascinated by that because it was the only thing about computers that also seemed like mathematics.
>
> I wanted to find a nice way to define semantics that would suit the Backus-Naur form syntax. Attribute grammars were a natural development because they matched intuitive notions of meaning.

Before Knuth started this work, the most economical description of the meaning of a programming language was its multi-thousand-line translator! This stood in sharp contrast to the elegant economy of the Backus-Naur description of its syntax. Knuth figured out a way to associate rules of interpretation (what he called *attribution rules*) with rules of the grammar.

For example, in an algebraic expression like $x + y/z$, an attribute grammar attaches a call to a division instruction at the parse tree node of y/z, and a call to an addition at the parse tree node of $x + y/z$ (see Figure 1). Thus, the meaning of a computer program (in this case, add x to the quotient y/z) is built up or "synthesized" from its component parts. Programming language theorist Ned Irons had proposed this idea in 1960 for simple languages, but synthesis, Knuth knew, was not enough. Programs in complicated languages carried contextual information around that had to be understood to permit a precise interpretation of the meaning of the names in a program. In a conversation in 1967, Peter Wegner suggested conveying information down a parse tree as well as synthesizing from the bottom up. After first thinking that this was a preposterous idea, Knuth figured out a way to use it and "inherited" attributes were born. Much of the technology of modern compilers dates from these insights.

A Precise Analysis

Since his work on compilation, Knuth has worked on algorithms—efficient methods for computers to accomplish frequently needed tasks. Among the most fundamental algorithms are those for sorting numbers or names in a list to put them in order or searching lists to find the addresses associated with a name. Knuth devoted a volume of *The Art of Computer Programming* to just these tasks. But his main contribution in this area comes from his "exact analysis" of algorithms.

When other computer scientists might be satisfied to say that an algorithm takes time proportional to the square of its input, Knuth would prove that it takes exactly 3.65 times the square of the input.

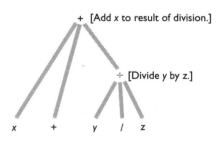

+ [Add x to result of division.]

÷ [Divide y by z.]

x   +   y   /   z

Figure 1   Parse tree for $x + y/z$. Synthesized attributes (in brackets) convey the meaning of the expression from the bottom to the top of the parse tree.

The way I do analysis of algorithms is different, I guess, from the vast majority of workers in the field. The most unique part of my work is to be able to say that something is 10 to 15 percent better than something else. It's a matter of taste and temperament. I'm better at seeing the small picture, from my training—others are better at seeing the large picture.

Some scientists are like explorers who go out and plant the flag in new territory; others irrigate and fertilize the land and give it laws and structure.

But Knuth the lawgiver has done much significant exploration even beyond his work on compilers. For example, collaborating with Peter Bendix, one of his undergraduate students at CalTech, Knuth created an algorithm for exploring the consequences of mathematical axioms.

Mathematicians learn axioms for an abstract structure called a group; then they prove many consequences of those axioms. One day I stumbled on the fact that a computer can be made to discover such proofs systematically. You start out with three axioms for groups and derive seven consequences; then you can demonstrate that these ten things are "complete"; every identity that anybody can derive from the original three can be derived from the ten by a very simple procedure.

The basic idea of the algorithm is to start with axioms in the form of equations and to consider them as "reductions." For example, the axioms may be $a \times (b \times c) = (a \times b) \times c$ and $a \times 1 = a$. These generate the reductions $a \times (b \times c) \to (a \times b) \times c$ and $a \times 1 \to a$. A typical conclusion is that $a \times (1 \times b) = a \times (b \times 1)$. The underlying idea is to show that two expressions x and y are equal by showing they both reduce to some third "canonical" expression z. The algorithm finds z by applying reductions to x in any order, until no more reductions apply, and similarly for y. In the above example, $a \times (1 \times b) \to (a \times 1) \times b \to a \times b$ and similarly $a \times (b \times 1) \to a \times b$. The Knuth-Bendix algorithm will generate new reductions if the original set is incomplete; for example, the additional axiom $a \times a' = 1$ would lead automatically to reductions such as $(a \times b)' \to b' \times a'$.

Scientists from physicists to political scientists invent axiom systems all the time. The Knuth-Bendix algorithm can help explore the

implications of those axioms and verify proofs which use them—
often more accurately than the scientists' own methods.

In 1968, Knuth moved to Stanford University, which had become
one of the top three computer science departments in the world (with
M.I.T. and Carnegie Mellon). With graduate student Vaughan Pratt,
he discovered a simple, yet extremely efficient way to search texts for
a string of characters. A similar method was discovered at about the
same time by James Morris, so it is now called the Knuth-Morris-
Pratt algorithm.

Consider the problem of finding the string "init" in a long text. The
obvious way is to look at the characters of text successively and, if you
find an i, look at the next character and, if that's an n, look at the character after it, and so on. The tricky question is what to do if you find
matches for a while and then find a mismatch. For example, if the text
is "isininity . . . " then you will begin matching at the first letter of the
text, but then detect a mismatch at the second letter. You will start over
at the third letter and continue matching at the fourth and fifth letters,
but will detect a mismatch at the sixth. At that point, it would be a mistake to just continue with the seventh letter in the text looking for the
first i in init, because you would find "ity . . . " and therefore would
falsely conclude that "init" is nowhere in the string. You might solve
this problem by starting the comparison over with the fourth letter in
the text, since you had failed to find a match starting at the third letter.
The trouble is that you would then look at the fourth, fifth, and sixth
letters of the text twice. This is inefficient, particularly for longer patterns, because you might look at many characters repeatedly.

Knuth, Morris, and Pratt sought an algorithm that was both correct and efficient. They drew inspiration from ideas developed by
Robert Boyer and G. Strother Moore and advances in automata theory pioneered by Steve Cook.

> Our paper was actually going to be called "Automata Theory Can
> Be Useful," because the method is not an algorithm that a pro-
> grammer would normally think of. By knowing automata theory,
> we had important clues about how to proceed. [The final title was
> more straightforward: "Fast Pattern Matching in Strings."]

The algorithm builds a table that makes it emulate a simplified
version of a finite state automaton or finite state machine (see the

chapter on Rabin, page 68). These are theoretical machines having states and transitions. Some states are called starting states (where the machine starts processing) and some states are called accepting states (where the machine declares that it has found a pattern). In this simplified version, the states correspond to the number of string characters that have been matched. Thus, for the four character pattern "init," we start in state 0 and declare a match in state 4. Transitions describe the progression from one state to the other depending on the next input character. Here is what the machine does when searching for the pattern "init" in the string "isininity."

CHARACTER	RESULTING STATE
i	1
s	0
i	1
n	2
i	3
n	2
i	3
t	4,match
y	0

The machine advances to state 1 because the first character is an "i," but returns to state 0 because the second character is an "s." Most algorithms would do this. The key insight of the Knuth-Morris-Pratt algorithm is that failing to find "t" after the first "ini" does not necessitate backing up to reconsider previous characters. Thus the algorithm avoids reconsidering previous characters, a significant practical as well as theoretical advantage.

Fonts

Throughout his life, Knuth had been intrigued by the mechanics of printing and graphics. As a boy at a Wisconsin summer camp in the 1940s, he wrote a guide to plants and illustrated the flowers with a stylus on the blue ditto paper that was commonly used in printing at that time. In college, he recalls admiring the typeface (Modern) used in his math textbooks. But he was content to leave the mechanics of designing and setting type to the experts.

I never thought I would have any control over printing. Printing was done by typographers—hot lead, scary stuff. Then in 1977, I learned about new printing machines that print characters made out of zeros and ones—just bits, no lead. Suddenly, printing was a computer science problem. I couldn't resist the challenge of developing computer tools using the new technology with which to write my next books.

Knuth put his other projects on hold for what turned out to be nine years, while he designed and implemented two computer languages for digital typography. The first, called TEX, positions letters and other symbols on the page. The other, called METAFONT, defines the shapes of the letters themselves. These programs are now available worldwide as free software. They have more than a million users including the publisher of this book.

Knuth explored the field of typography with characteristic thoroughness. For example, he wrote a paper called "The Letter S" in which he dissects the mathematical shape of that letter through the ages, and explains his several day effort to find the equation that yields the most pleasing outline.

Compilers, algorithmic analysis, fonts—is there a common thread among his wide-ranging accomplishments? Knuth asserts there is.

It's not true that necessity is the only mother or father of invention. The other part is that a person has to have the right background for the problem. I don't just go around working on every problem that I see. The ones I solve, I say, "Oh, man, I have a unique background that might let me solve it—it's my destiny, my responsibility."

His unique background includes a love of language and grammar, a clearheaded and vast knowledge of mathematics, a strong visual aesthetic, a will to understand, and a love for programming. To Knuth, the last two are closely related.

In general, whatever you're trying to learn, if you can imagine trying to explain it to a computer, then you learn what you don't know about the subject. It helps you ask the right questions. It's the ultimate test of what you know.

For example, music theory was developed in order to have an objective rather than a subjective answer to what sounds good and what doesn't. We know Mozart sounds good because of the harmony, etc. But you learn how incomplete those rules are when you think about writing a program for a computer, trying to get a machine to create really good music.

Now a youthful professor emeritus at Stanford, Knuth spends most of his time writing the combinatorics volume of *The Art of Computer Programming*. But there are other temptations—the pipe organ beckons to him to write more music. His literary interests may compel him to write another novel. (His 1974 book *Surreal Numbers* is about two college dropouts who develop a mathematical system.) Conferences and consulting projects worldwide divert his attention. Aside from talent, enthusiasm, and legendary work habits, there is no secret to his vast accomplishments.

> I do one thing at a time. This is what computer scientists call batch processing—the alternative is swapping in and out. I don't swap in and out.

HOW LEGENDS ARE BORN

Folklore holds that Knuth is the greatest computer programmer of all time. Consider the following anecdote from Alan Kay.

> When I was at Stanford with the AI project [in the late 1960s] one of the things we used to do every Thanksgiving is have a programming contest with people on research projects in the Bay area. The prize I think was a turkey.
>
> McCarthy used to make up the problems. The one year that Knuth entered this, he won both the fastest time getting the program running and he also won the fastest execution of the algorithm. He did it on the worst system with remote batch called the Wilbur system. And he basically beat the shit out of everyone.
>
> And they asked him, "How could you possibly do this?" And he answered, "When I learned how to program, you were lucky if you got five minutes with the machine a day. If you wanted to get the program going, it just had to be written right. So people just learned to program like it was carving in stone. You sort of have to sidle up to it. That's how I learned to program."

Robert E. Tarjan

IN SEARCH
OF GOOD STRUCTURE

I visualize structures, graphs, data structures.
It seems to come easier than a lot of other things.

—ROBERT E. TARJAN

here are three marks of a good idea in computer
science. It's used often in practice; it's taught early
to young computer scientists; and it suggests av-
enues for research. By these criteria, Robert Endre Tarjan has come
up with many good ideas.

His Ph.D. work with John Hopcroft at Stanford University on planarity testing and other graph algorithms has spawned applications
ranging from more efficient chip layout to better map layouts. These
algorithms underscored the power of a programming technique
known as *depth-first search,* which is taught to every undergraduate.
Depth-first search is also used billions of times a day in game, spreadsheet, and graphics programs. Tarjan's later work on efficient network
flows with Dan Sleator and Andrew Goldberg has helped designers
figure out how much capacity to give each link in networks to improve
flows of everything from offshore oil to telephone calls. His work on
a pair of simple data structures known as up-trees and splay-trees introduced new techniques for measuring the efficiency of algorithms.

Other work with Neil Sarnak, Sleator, and James Driscoll suggested a data structure that can hold information about the past as well as the present, in a very efficient form. Such "persistent" data structures enjoy increasing use today in robotics and database systems.

Tarjan has also brought about a cultural change in the study of computer science: get the right "big idea" and then create the data structures to support it most efficiently. This strategy has led him and others to squeeze significant improvements out of Dijkstra's shortest path algorithm and other fundamental algorithms.

His work is characterized by minimalism, elegance, and an implicit reach toward generality.

> A good idea has a way of becoming simpler and solving problems other than that for which it was intended.

Robert Endre Tarjan was born in Pomona, California, in 1948. From an early age, he was interested in science.

> In about the seventh grade, I got interested in math from reading Martin Gardner's columns in *Scientific American*—playing games and solving puzzles. Before that, I was interested in astronomy. I wanted to be the first person on Mars.

Tarjan's father, a child psychiatrist specializing in mental retardation, ran a state hospital. Tarjan got a job there when he was in junior high school, and worked with IBM card punch collators— "precomputers" as Tarjan calls them. At a summer science program following his junior year of high school in 1964, Tarjan worked with real computers for the first time.

> The idea was to take observations of an asteroid and calculate the orbit. We had a chance to use a computer at UCLA so I learned Fortran.
>
> By that time, the state hospital had gotten a tiny computer, too. So I worked on that during summers off from Caltech doing programming—statistical studies of the patient population.

At CalTech, Tarjan majored in mathematics, but took all the graduate computer science courses as well.

I wanted to do computer science because it was a more applied kind of mathematics. I had an interest in artificial intelligence, mostly from a logic and theorem-proving perspective. But when I got to Stanford and started taking the AI courses, I decided it was pretty fuzzy stuff.

Stanford boasted some of the leading minds of computer science including Donald Knuth and John McCarthy.

Knuth was first of all inspiring. Inspiring in the sense that his focus has been very concrete analysis. He has always been interested in mathematical rigor—getting exact results.

Another inspiration was John Hopcroft, on sabbatical at Stanford from Cornell. He arrived during the summer before Tarjan's second year in graduate school and moved into an adjoining office. The two soon began a collaboration that ultimately led to the Turing Award in 1986.

I had taken McCarthy's symbolic processing course which was mostly a course in Lisp. He suggested that the students write programs to determine whether a graph is planar.

I didn't have any expectations—I was just a graduate student trying to solve interesting problems.

Can a Graph Be Laid Flat?

A graph is a set of nodes and edges between nodes. (In the graph in Figure 1, the nodes are small circles and the edges are the connecting lines.) Graphs are used to represent many different real-world phenomena. For example, nodes may represent molecules and edges may represent molecular bonds; nodes may represent people and edges may represent mutual acquaintances, or nodes may represent intersections and edges may represent streets.

For certain applications, it is important that the edges not overlap. For example, in a circuit layout an overlap of the edges (wires) would cause a short circuit. A natural question, then, is whether a graph can be laid out so its edges don't overlap while its connections are preserved. Such a graph is called *planar*. In Figure 1 the edges in

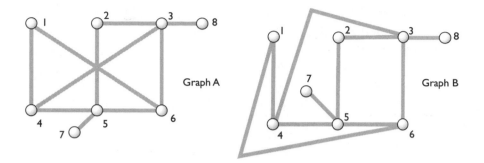

*Figure 1   A planar graph. Graph A can be redrawn so that its edges don't over-
lap but the connections are still preserved, as shown by graph B.*

graph A do overlap, but we can lay out the graph so the edges don't
overlap, as in graph B.

Tests to determine whether a graph is planar go as far back as
Euler in the eighteenth century. The Swiss mathematician showed
that if the number of nodes is N, then no graph with $3N - 6$ or more
edges can possibly be planar provided N is at least 3.

In 1930 the Polish mathematician Kuratowski showed that every
nonplanar graph *must* contain a graph with the connectivity relationship of one of the graphs in Figure 2. McCarthy had suggested

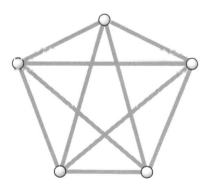

Complete graph on five vertices

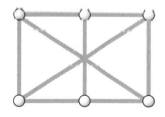

Complete bipartite graph

*Figure 2   Every nonplanar graph must have a structure (subgraph) like one of
those shown here.*

that students use the Kuratowski condition in their programs. Tarjan quickly decided that the resulting algorithm would be too inefficient. The Kuratowski test *can* be converted to an algorithm, but the time required is proportional to the sixth power of the number of vertices. This would imply a trillion steps for a graph with about 100 vertices.

> So, I was already thinking about planarity testing. When Hopcroft arrived, we started talking about algorithms and efficiency.
>
> Hopcroft came up with an algorithm for finding biconnected components which was sort of a loose sketch to begin with. But what it really was was a depth-first search. I thought about it and tried to make the principle of what was going on in his algorithm more rigorous.

Applying Depth-First Search

Depth-first search had been widely used for finding solutions to problems in artificial intelligence, where it was called *backtracking*. Backtracking explores different approaches to solving a problem without repeating any one approach. In AI, it had been used in games to search systematically among sequences of moves and countermoves. Tarjan and Hopcroft used depth-first search as a systematic way to explore a graph. (In a depth-first search a single approach is followed to its conclusion before another one is tried. In another strategy, called *breadth-first,* many approaches to solving a problem are worked on at the same time. Which approach is used depends on the problem.)

Starting from a node *n* of the graph, depth-first search chooses an edge leaving *n*. Traversing the edge leads to a new node. In general, the program selects and traverses unexplored edges leading from the most recently visited node. Depth-first search ensures that no edge is traversed more than once.

Hopcroft and Tarjan used depth-first search to implement a strategy that had been pioneered by L. Auslander and S. V. Parter in 1961 and improved by A. J. Goldstein in 1963. In contrast to Kuratowski's test for nonplanarity, Auslander, Parter, and Goldstein had proposed an algorithm that attempted to lay a graph out on a plane directly.

Hopcroft and Tarjan's algorithm builds directly on top of Goldstein's. Its basic steps are the following:

1. Test that the graph doesn't have more edges than Euler's condition would allow (if it does, then it isn't planar).

2. Divide the graph into biconnected components. A *biconnected component* of a graph is a collection of nodes having the property that a path exists between any two nodes in the collection and there will still be a path even if any single node in the collection is removed. If you find this too abstract, think of it this way: your city is biconnected if it is impossible for road work on a single intersection to prevent you from driving to your favorite jazz club.

3. Use depth-first search to find a cycle in the graph. A *cycle* is a path in the graph that starts and ends at the same node.

4. Removal of the cycle breaks the graph into connected pieces, called *bridges*. For each individual bridge, test whether the bridge and the cycle together form a planar graph. This can be done by applying this same algorithm to each bridge. (The algorithm avoids an infinite regression in this "recursion" because the bridge is smaller than the original graph.)

5. If the cycle is laid out in the plane, each bridge must go either completely inside or completely outside the cycle. Certain pairs of bridges interfere with each other and must be assigned to opposite sides of the cycle.

The key to making this algorithm efficient is to do the entire computation, including the planarity tests of contained graphs (the official term is subgraph) using a single depth-first search. The details are technical, and their paper, "Efficient Planarity Testing," goes on for twenty pages in the extremely dense *Journal of the Association for Computing Machinery.* But you can see the basic construction in Figure 3, which shows a depth-first search of a planar graph that is a slightly simplified version of graph A in Figure 1. The solid arrows represent one possible approach to exploring the nodes in the graph and the dotted arrows represent graph edges unaccounted for by the solid arrows. For example, the edges from 4 to 1 and from 5 to 6 are

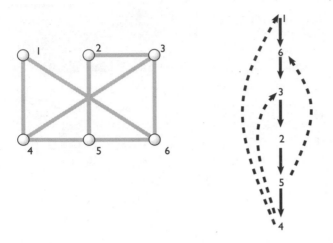

Figure 3   Depth-first-search planarity testing. A depth-first search of the graph on the left can yield many different trees. The tree shown by the solid edges in the diagram on the right is one possibility. The dotted edges correspond to edges which appear on the graph at the left but which are not part of this particular tree representation.

in the graph but are not directly in the tree, and so are shown as dotted edges. The fact that these extra edges, called *back edges,* point to ancestors in the tree is one of the many nice features about depth-first search that Tarjan and Hopcroft discovered.

For the graph in Figure 3, the cycle found in the second step of the algorithm would be $1 \rightarrow 6 \rightarrow 3 \rightarrow 2 \rightarrow 5 \rightarrow 4 \rightarrow 1$. The fifth (checking) step mentioned above consists in observing that the two remaining dashed edges, $4 \Rightarrow 3$ and $5 \Rightarrow 6$, can be embedded one inside and one outside the cycle on a plane. This means that the graph is planar.

Tarjan and Hopcroft's strategy reduces the problem of planarity to one of determining whether the depth-first search tree plus the back edges is planar. This can be done very efficiently.

In contrast to all previous approaches to this problem, the Hopcroft-Tarjan algorithm runs in linear time (that is, time proportional to the size of the graph). This means that doubling the size of the problem only doubles the time it takes to solve it. By contrast, using the Kuratowski criterion, doubling the size of the problem could increase the time by a factor of more than 60.

Planarity made a big splash because everybody thought it was a really hard problem so it should take a lot of time. When I implemented our planarity-testing algorithm, it ran very fast. It became clear that using sophisticated algorithms, one could solve certain problems a lot faster.

In the early 1970s this was a minority view. The mathematical study of the time cost to perform computations was only a dozen years old. It had started with Michael Rabin's work in 1959 and continued with work by Donald Knuth, Juris Hartmanis, Steve Cook, and Leonid Levin. But the idea that the cost of a method depended on the number of operations it took to solve a problem had not yet penetrated into the minds of practicing programmers.

Instead, practitioners and many academics ran their algorithms on their own computers and compared the running times with the times of published algorithms on older computers. This meant that a poor algorithm running on a fast computer would appear to be better than a good algorithm running on a slow computer!

In their planarity article, Hopcroft and Tarjan revealed their attitude toward contemporary practice with academic brutality: "Surprisingly little work has been directed towards a rigorous analysis of their running times, however, and algorithms continue to appear that are clearly inferior to previously published ones."

Hopcroft and Tarjan advocated, instead, measuring an algorithm's speed based on the number of basic operations it required—additions, comparisons, edge traversals, etc. Because of the practical success of the planarity algorithm, this approach soon became enshrined in the practice as well as the theory of computer science. It must have also helped that Al Aho of Bell Labs, Hopcroft, and Jeff Ullman at Princeton University wrote a widely adopted text on algorithms in 1974 in which they measured algorithmic speed in this way.

The other offshoot of efficient planarity testing was the recognition that depth-first search itself had widespread applications. At the ceremony in which Hopcroft and Tarjan received their Turing Award, the winner of the year's best computer chess program noted that his program had used depth-first search over 40 million times during the chess match.

After receiving his Ph.D. in 1971, Tarjan went to Cornell as an assistant professor. His interest in algorithmic efficiency continued.

Union-Find and Amortization

Soon after arriving at Cornell, Tarjan began work on a deceptively simple problem called *union-find*. Its efficient implementation would speed up the solution to a wide variety of other problems.

In many graph algorithms, nodes must be separated into different collections known as *partitions*. During the course of an algorithm, different partitions may merge to form larger partitions.

For example, suppose a problem specifies that all the nodes in a single partition must be connected—there must be a path from every node to every other node in the partition.

Suppose a step of the algorithm to solve the problem discovers that some node in partition A is connected by an edge to some node in partition B. Then the two partitions must be merged into one whose members are the set union of the members of partition A and the members of partition B.

Figure 4 shows an example of this. Initially, each element is an isolated set by itself (see Figure 4, part A). In the example, partition A is the set {1, 2, 4} and partition B is the set {3, 6} and the resulting partition has members {1, 2, 4, 3, 6}.

Each time two sets are merged together, an arrow is drawn from the "canonical node" of one to the canonical node of the other, usually from the smaller set to the larger one (the dotted edge from 6 to 2 in Figure 4, part B). The *canonical node* of a set is the node that has no edges leaving it. This is the *union* operation of the union-find algorithm.

Since membership in partitions changes over time, a good data structure must make it easy for an algorithm to find out whether two nodes belong to the same partition at a particular time.

The structure in part B of Figure 4 is an example of an *up-tree*, a term proposed by B. A. Galler and M. J. Fischer in 1964 for a structure in which the partition identifier is the node in the partition that has no edges leaving it. Two nodes can discover whether they belong to the same partition by following the directed edges up to the parti-

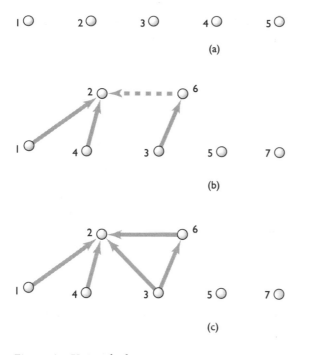

Figure 4   *Union-find.*

tion identifier. In other words, to find out if two nodes belong to the same set, one simply checks to see whether they have the same canonical node. This is the *find* operation of the union-find algorithm. Thus find(3) = 2 and find(1) = 2 so they are in the same set, whereas find(5) = 5 (since it has no edges leaving it), so 5 is in a different set from 6 and 3.

Each find operation shortens the path to the partition identifier as much as possible. It does this extra work, known as *path compression,* to speed up future union and find operations. Figure 4, part C, shows how the Hopcroft-Ullman algorithm uses path compression to shorten the path from 3 to 2. In this way, future finds from 3 would require only one step.

The role of union-find in many algorithms is like that of an elevator in a skyscraper. One can construct the building without a fast elevator, but people won't be happy. For this reason, it was essential to find the fastest union-find algorithm possible. To do that required a very careful timing analysis. Tarjan's principal contribution was to do this analysis.

Earlier, Hopcroft and Ullman had supposedly proved a linear upper time bound. As in planarity, a linear upper time bound means that doubling the size of the problem doubles the time. At an IBM workshop on algorithms in 1973, Mike Fischer of Yale described a bug in the Hopcroft-Ullman proof. Tarjan started thinking about the problem. He wondered whether there was some kind of construction of bad examples which showed that the time bound was in fact superlinear, that is, would require more than twice the time for a problem twice the size. Tarjan came up with a doubly recursive construction which gave an inverse Ackermann's function as a lower bound for the problem, implying that no algorithm could be linear. Later, he proved that the lower bound was also an upper bound by demonstrating an algorithm that took this amount of time.

> Thus, this rather esoteric function appears in an inherent way in the analysis of this very simple data structure. That was the amazing thing about this problem. There was no way to predict that. Most people believed that the running time of the algorithm was linear.

This belief was in fact close to the truth. The Ackermann function was originally invented as an illustration in an advanced branch of mathematics known as recursion theory. It grows extraordinarily fast. This means that the inverse Ackermann function grows extraordinarily slowly—so slowly that it will never be above 5 for any remotely practical application.

But as in planarity, the result was surprising—no one had ever thought the Ackermann function would have real life applications. This led researchers to look at the new analysis technique Tarjan had invented.

Analyzing union-find is hard because the find operation will require just one pointer traversal once the path has been compressed (Figure 4, part C), but may require a long traversal if the path hasn't been compressed.

A single find operation can take many operations. In analytic techniques up to that time (including those used by Hopcroft and Tarjan for planarity), researchers would measure the difficulty of an algorithm by multiplying the number of operations by the worst case time of any single one. This admittedly conservative method usually worked well. Tarjan saw that this standard analytic technique would give a misleadingly large time for this problem, however.

While a single find operation might take a long time, it could reduce the time for future find operations by performing path compression along the way. In terminology that Tarjan and his student Danny Sleator would borrow from the field of accounting a few years later, the extra work of one find operation would be *amortized* over the many find operations that would benefit from it.

> The point is you've got a long sequence of operations on a data structure. You're not interested in individual operations, but in the average time of an operation over the sequence.

Just as depth-first search has become a standard algorithmic technique, amortization has become a standard analytic technique. This has given rise to many algorithms in which one operation behaves altruistically with respect to its fellow operations.

Maximum Network Flow

In 1973 a second harsh Ithaca winter spurred Tarjan to accept an offer from Berkeley, where he stayed for two years. In 1975, Tarjan returned to Stanford where one of his graduate students was Danny Sleator. The idea of amortized time would reemerge when he and Sleator worked on the so-called maximum network flow problem.

> You're given a directed graph with capacities on the edges and a source and a sink. Imagine the edges are pipes and you're pumping stuff through the pipes and you want to get the maximum amount of stuff from the source to the sink.

Graph A of Figure 5 is a graph representing an oil network each of whose pipes (edges) has a capacity (indicated by a number), which is the maximum flow that can move through that pipe (edge). Graph B shows the maximum flow through the network given those capacities and a total flow of *s* to *t* of 115. The network flow problem consists of discovering the maximum through a given network.

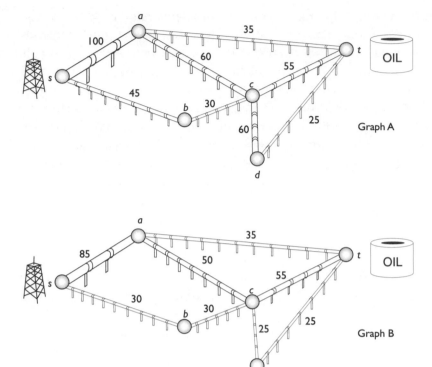

Figure 5   Maximum flow of a network.

In 1956, L. R. Ford and D. R. Fulkerson published the first algorithm to solve the problem.

1. Start with a flow of zero and call that the current flow. This
   clearly respects the capacities (i.e., no edge is oversaturated).
2. Attempt to find a path, an *augmenting path,* from the
   source to the destination in which every edge has some extra
   capacity. Force as much flow as possible, without oversatu-
   rating any edge, through that path. Add the flow through
   the augmenting path to the current flow. Repeat step 2 until
   no more augmenting paths can be found.

As Ford and Fulkerson themselves pointed out, the algorithm
was inefficient for certain examples. Worse still, it didn't reach the
right answers if the capacities were irrational. More than a decade

later in 1969, J. Edmonds and R. M. Karp proposed a refinement to the algorithm that resolved the problem and made it much more efficient: choose an augmenting path having the fewest edges. This gave an algorithm that ran in time proportional to the number of nodes times the square of the number of edges.

Working independently, the Soviet mathematician E. A. Dinic in 1970 made the same discovery as Edmonds and Karp. But Dinic carried it one step further by finding a way to locate all augmenting paths having the same number at once. Subsequent algorithms proposed by Indian, Israeli, Soviet, and American researchers built on this idea. Tarjan and Sleator's algorithm achieved a faster algorithm through the invention of a new data structure.

> It turned out that the crux of the matter was to get an efficient data structure for solving one piece of the problem [pushing edges to saturation]. The data structure, known as a dynamic tree, became the basis for Sleator's doctoral thesis.

In 1980, Tarjan and Sleator went to Bell Labs in Murray Hill, New Jersey. They resumed work on dynamic trees.

> We eventually figured out that if we didn't concentrate on the worst case, but were willing to do this kind of time averaging [amortization], we could really simplify this data structure.
>
> We started thinking about the notion of amortization in a more systematic way. We came up with the idea of a self-adjusting data structure, which you design not to be efficient in the worst case but to be efficient in the amortized case. We created a self-adjusting data structure we called a *splay tree*, which has very nice properties.

Like path compression, splaying a tree is a task that one operation performs for the benefit of future operations on the tree. The splaying itself may be expensive however, so Tarjan and Sleator had to use amortization techniques to show that the benefit was worth the cost.

Competitiveness

Planarity, network flow, and the other problems Tarjan had looked at until the early 1980s all had definite right answers. A graph is either planar or it is not; a flow is either maximum or it is not. By contrast, certain problems involve only a comparative notion of merit, and do not lead to a definitive right or wrong solution.

An analogy may help. We believe an investment newsletter writer is good if he suggests good stocks. One possible measure of quality is how well an investor does by following the newsletter writer's advice compared with following the advice of a mythical clairvoyant advisor whose knowledge of the future is perfect. Even though a clairvoyant advisor is an impossibility, she provides a good standard of comparison. Tarjan and Sleator proposed exactly such a standard of comparison for the fundamental problem of buffer management.

A computer has fast memory called *RAM* (random access memory) and another kind of memory, 1000 times slower, called *disk*. Since RAM cannot normally hold all the data of interest to the user, the goal of buffer management is to keep in the RAM buffer data that is likely to be used soon. Since no algorithm knows the future, real algorithms try to achieve this goal by making guesses based on history. The most popular such algorithm is called *least recently used*.

> You keep track, for each page, of the time it was most recently looked at. The oldest page is the one that is thrown away.

The hypothesis underlying least recently used is that the pages your programs touched most recently are the ones they are likely to touch again soon. The past is a mirror of the future.

> Suppose you know the future—you know the entire sequence of page requests. Then how can you construct the best possible paging strategy? This was done by L.A. Belady at IBM in the 1960s. The algorithm kicks out the page whose next use is the farthest in the future. But you can't implement this off-line [clairvoyant] strategy, because you don't know the future. The question this raises is how well can you do with a simple, on-line strategy compared to the best possible off-line strategy?
>
> Sleator coined the word "competitive." An on-line algorithm is competitive when its performance is within a constant factor of the best possible off-line algorithm.

Tarjan and Sleator showed that the least recently used algorithm will do reasonably well by this criterion. For a buffer of size S, least recently used will kick out at most twice as many pages as a clairvoyant algorithm would if the clairvoyant algorithm had a buffer of size $S/2$. Besides being important in its own right, this work put the measurement of on-line algorithms for scheduling, resource distribution, and related problems on a rational basis.

Persistent Data Structure

While at Bell Labs in the early 80s, Tarjan taught at New York University as an adjunct professor. With NYU graduate student Neil Sarnak, Sleator, and Carnegie Mellon student Jimmy Driscoll, he began to look at long-lived data structures.

> The problem was to construct a data structure in which you can keep track of previous versions as well as the most recent version. And to do it efficiently, without copying the entire data structure.

A *data structure* is a graph structure stored in a computer's memory to make access to data faster. Typically, data structures are trees, which permit programs to get to the data they need fast. For example, the most widely used data structure in database systems, the B-tree, invented by the German computer scientist Rudolf Bayer and the American E. M. McCreight while at Boeing in the early 1970s, can find a requested fact among 100 billion others in less than 0.01 second. This is roughly 10,000 times faster than if each fact had to be searched one at a time. Tarjan and his colleagues called their long-lived data structure a *persistent data structure*. Persistent data structures are nearly as fast normal ones for current data; in addition, they enable programs (and therefore users) to access the state of the data in the past as well at little extra cost.

Figure 6 shows an example of a persistent data structure proposed by Tarjan and Sarnak. If the data represented by node X in the figure must be updated, then the most efficient way to do so is to replace X by the new value X'. The trouble with that approach is that it is no longer possible to recall the state of the database at time 1. Using the Tarjan-Sarnak method, the program can obtain the

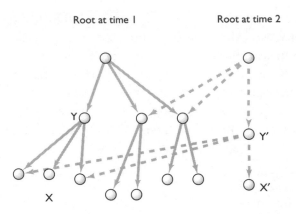

Root at time I   Root at time 2

Figure 6   Persistent data structure of Tarjan and Sarnak.

state at time 1 by starting at the root at time 1 and the state at time 2 by starting at the root at time 2. Since most of the data hasn't changed between time 1 and time 2, the data structure for time 2 holds only the differences. Instead of replacing X by X′ and losing history, the program constructs a new node to hold X′ and then new nodes for all ancestors of X′. For most structures this is not much more work than locating X and replacing X by X′.

This idea has had applications in computational geometry and parallel processing, but its main long-term application may be so-called *temporal databases*. These databases are designed to recreate snapshots of the past quickly and efficiently.

A Pattern in the Work

Tarjan sees a pattern in the many algorithmic problems he has selected.

> Half the battle is in choosing the problems—not in coming up with the solutions. If you get the right problems, if you ask the right questions, you're a long way to the solution.
>
> Find problems that emerge out of applications. The danger is to get into some little branch of theory that becomes self-feeding and doesn't tie back into the real world. I have always tried to

work on problems that have some practical importance, for which I thought I could get algorithms that would be practical.

[On the other hand], you can never tell when some idea generated in some isolated area turns out to connect with something else. The magic of mathematics and theoretical computer science is all the unexpected connections. You start looking for general principles and then mysterious connections emerge. Nobody can say why this is.

Tarjan sees a social pattern to his successes as well.

Interaction is very important and having cooperative spirit. One of the greatest joys of being a professor is having graduate students. It gives you tremendous leverage in doing research and also it's very motivating because you've got people whose minds are fresh and open and eager to learn.

In the end, however, research is largely an individual effort. Even for Tarjan, days can go by without progress.

What do you need to be successful? You need brains but you also need stick-to-itiveness. Many tries at a solution can fail, but then on the last try something magical happens.

Leslie Lamport

OF TIME, SPACE, AND COMPUTATION

A distributed system is one in which the failure of a computer you didn't even know existed can render your own computer unusable.

—LESLIE LAMPORT

reat conceptual advances sometimes stem from re-considering commonsense assumptions. In special relativity, Einstein reached his famous conclusions about mass and energy by starting with the hypothesis that light travels at the same speed with respect to every frame of reference. On the way, he showed that there is no such thing as simultaneous events. Two events that appear to be simultaneous in one frame of reference may not appear to be simultaneous in another. Einstein thus purged physics of the commonsense notion that simultaneity carries a meaning independent of the observer. Instead, he asserted that the relative order in which two events occur depends upon the observer.

In computer science, Leslie Lamport has caused the designers of distributed computing systems, collections of computers connected by networks, to rethink their assumptions. Inspired by what Einstein did in physics, Lamport showed that the relative order of events in a distributed system can depend on the observer. Relative order is based on the sequence of locally related events and the sending and receipt of messages.

Born in New York City in 1941, Leslie Lamport is the son of immigrants from eastern Europe. The Depression prevented Lamport's father from realizing his dream of becoming a doctor, and he ended up in the dry cleaning business instead.

> Perhaps being the son of immigrants made me more competitive. Perhaps it made me more self-reliant—almost all my early work was done without collaborators, pretty much in isolation.

Lamport's intense curiosity about math led him early on to test the limits of his teachers' competence. One day, a junior high school teacher was showing the class how to factor polynomials.

> I said, "How do you know there's only one way to factor a polynomial?" You'd think if a kid asked a math teacher a question like that, his eyes would light up and he'd talk about unique factorization and stuff like that. All my teacher said was, 'It just is.' He couldn't provide an explanation.

At the Bronx High School of Science, New York City's elite public high school for science and math students, Lamport was more impressed with his teachers, though he remained hard to please.

> I had a good calculus class at Bronx Science that actually gave rigorous definitions and proved things. I remember one thing that wasn't proved—that the derivative of the natural logarithm of x is $1/x$. I remember feeling very unsatisfied with that.

Not only did Lamport want to see proofs, a rare desire even among those who enjoy high school mathematics, but he also learned not to trust any proof presented to him.

> I got interested in non-Euclidean geometry and spent a lot of time at the New York Public Library, on Forty-Second Street. At some point, I read a book with deliberately fallacious proofs of theorems like "all triangles are isosceles." That must have planted a seed in my mind that caused me to be suspicious of claimed proofs.

In 1956, Lamport's junior year in high school, he encountered his first computer at the IBM building in New York. It was the same year Backus introduced Fortran, the first high-level programming language. (See the chapter on Backus.)

> I remember visiting IBM in New York and getting some tubes from their computers that were still working but were marginal. Then a friend and I put some flip-flops [an elementary memory circuit] together and made a one-digit counter.
>
> The real connection with computers came the summer before I entered M.I.T. There was some program for bright students sponsored by the City of New York, so I wound up working for Con Edison in a ridiculously boring position with the keepers of the blueprints. If they had to dig up the streets, they had to find the blueprints that showed where the wires were.
>
> But there were computers at Con Edison. I hung around them and got myself transferred to the computer department.

Lamport read the manuals for the IBM 705 and started to program when the computers were left idle during lunch breaks. His most ambitious program computed the first 256 digits of the mathematical constant e, a remarkable number that plays a central role in calculus, especially that part of calculus which Lamport had wanted to see proved.

> I remember sitting there at lunch and somebody coming in—some grownup commenting, "He's playing with it just like it's a toy." He was right. It was a wonderful toy. And these grownups didn't appreciate what a wonderful toy it was.
>
> I've never done interesting research by sitting down and deciding that X is a Very Important Problem, and that I should therefore solve X. I doubt if anyone else has, either. Good research comes from thinking that something is fun, and playing with it.

After graduating from Bronx Science, Lamport went on to M.I.T., ostensibly to become an engineer.

> As soon as I got to M.I.T., I decided engineering wasn't for me. I didn't know what engineering was about, but it didn't seem appealing. My parents didn't care much for my decision, since being a mathematician didn't seem to pay as much as being an engineer. But I

would have an opportunity to become a college professor, so that was respectable enough.

Math was in a horrible state at that time. The feeling was that you didn't want to dirty your hands with anything applied. That dates back to [Godfrey] Hardy—the notion that pure math is superior to applied math. It permeated the teaching. If you could find a more abstract, more general way of doing something, that was marvelous because you got further away from the grotty reality.

One can speculate about the reasons for the Platonic approach to mathematics that Lamport encountered. Perhaps calculation was simply too laborious for mathematicians who preferred the glamour of theoretical exploration. A second reason—our speculation, not Lamport's—may have been that mathematicians, like some physicists, viewed the Manhattan Project as a tale with a moral: bringing calculations to their numerical conclusions can result in terrible new weapons. We have seen evidence for the second speculation in the laments of colleagues at NYU's Courant Institute whose analysis of stellar fusion led to better design techniques for firing mechanisms of hydrogen bombs.

In any case, an academic bias for theory over calculation was not unique to the United States. The history of Soviet mathematics includes many outstanding advances in pure mathematics, but very few in applied mathematics. Japanese mathematicians, by contrast, have historically been comfortable with calculation.

After obtaining his undergraduate degree from M.I.T., Lamport began his doctorate in mathematics at Brandeis University, where Dick Palais taught the first-year analysis course.

I discovered that you really could do analysis rigorously. I had this innate sympathy with his method of doing analysis. It fit in with my notion of wanting to see everything proved.

But Lamport's positive experience in Palais's course didn't prevent him from questioning school and mathematics. Concluding in his second year that math was pointless because of its distance from reality, he left. That was 1962. After taking up the piano and trying composition, Lamport considered returning to graduate school to study music. Instead, he decided to teach math at Marlboro College, a small liberal arts school in Vermont.

> I taught everything—I was the math department. There were some students who were very bright, but who couldn't handle the bull-shit of conventional schools.

Lamport's concern with the structure of proofs continued to haunt him.

> I tried to teach my students how to write a proof, but I failed. I realize now that the reason I was unsuccessful was that I was under the delusion that there should be a rigorous and precise way mathematical proofs are written, when in fact mathematical proofs are a literary form and aren't really rigorous. That's something I'm interested in changing.[1]

As teaching grew less attractive, graduate school became more so, especially since he began to have research ideas in physics—especially relativity.

> What really inspired me was a desire to understand time. People often say that time is just another dimension in the four-dimensional world of space-time. But it's not just another dimension, it's a very special one. In ordinary three-space, the distance between the origin $(0, 0, 0)$ and the point (x, y, z) is $\sqrt{x^2+y^2+z^2}$. But in four-dimensional space-time, the distance between $(0, 0, 0, 0)$ and (x, y, z, t) is $\sqrt{x^2+y^2+z^2-t^2}$.
>
> That minus sign is what makes the time direction different from the three space directions. I wanted to understand where that minus sign came from. I wanted to know why time was different from space.

An M.I.T. professor dissuaded Lamport from pursuing his ideas in relativity, but encouraged him to enroll in the graduate math program.

---

[1]Lamport is aware of how unusual his style is. In a recent paper, he wrote: "The proofs have been carried out to an excruciating level of detail. . . . The reader may feel that we have given long, tedious proofs of obvious assertions. However, what he has not seen are the many equally obvious assertions that we discovered to be wrong only by trying to write similarly long, tedious proofs."

To support himself, Lamport worked part-time for Massachusetts Computer Associates (COMPASS), which had a contract to write a Fortran compiler for the ILLIAC IV.

The ILLIAC was designed for a minimum of sixty-four processors, an enormous number for the 1970s. The goal was to use all of them together to solve massive numerical computations by executing instructions simultaneously on the various processors. The problem was how to take a Fortran program which does everything sequentially—one instruction at a time—and compile it to do multiple instructions at once. Lamport plunged into the project.

> That was the leading edge of computers at that time [1972]. I applied a little bit of simple mathematics to figure out what the algorithms were. I thought it was very straightforward stuff. It involved linear algebra. The reaction at COMPASS was like I was Moses coming down from the mountain with this very obscure sacred text that they would study. They eventually figured out what it meant in terms of an algorithm.

After receiving his Ph.D. degree in 1972, Lamport continued to work on the ILLIAC in COMPASS's Palo Alto office.

> I went off to California by myself, working long distance for COMPASS, but not talking regularly to anyone at COMPASS. I think that my isolation from the academic research community was a key to my work.
>
> I was outside academia so I wasn't influenced by the current fashion of what one was supposed to do.

The Bakery Algorithm

Lamport began to subscribe to the professional computer science literature, published mainly by the Association for Computer Machinery (ACM). An ongoing topic of discussion in *Communications of the ACM* (CACM) was Dijkstra's solution of the mutual exclusion problem (11 years old by this time).

Multiple computers tapping into a common resource in the computer memory must coordinate their access so that only one computer manipulates the resource at one time. This coordination is known as mutual exclusion: When one computer uses the resource, it should exclude the others from using the same resource.

As you may recall, Dijkstra formulated this problem and a solution by using a train routing analogy. If two trains from opposite directions must share a short piece of track, then to avoid collisions, only one train must occupy the shared track at one time. Engineers follow a system of semaphores to ensure that only one train at a time will be on the shared track, and one of Dijkstra's proposed solutions to the mutual exclusion problem used a programming mechanism called a "semaphore."

Semaphores permit a computer to gain exclusive permission to use a resource (such as a printer, a portion of a computer screen, or an internal program data structure). Semaphores are now supported directly by computer hardware.

Another Dijkstra solution, first published in *CACM* in 1965, did not require semaphores and therefore imposed less of a hardware burden. It had two drawbacks. First, a particularly unlucky computer might lose every race to the resource. In the technical jargon introduced by Dijkstra, that computer might "starve." In 1966 Knuth solved the starvation problem by a clever improvement to Dijkstra's algorithm.

But the Dijkstra and Knuth solutions had another drawback. They assumed that reads and writes to computer memory were "atomic." Atomicity in computer science is true to the original Greek meaning of the word: an atomic read or write is indivisible. If an atomic read and atomic write occur at about the same time, then either the read will occur completely before the write begins or vice versa. To achieve this effect, the hardware design must ensure that if several reads and writes arrive at a single memory location at the same time, they will be ordered in some way. Lamport's solution, the Bakery Algorithm, didn't put this burden on the hardware.

> I thought about the mutual exclusion problem and said how about the idea of everybody taking a ticket? I called it the Bakery Algorithm because, when I was growing up, the local bakery was the only store that used tickets to decide who would get served next. And of course, since there was no ticket distributor in the computer, everybody would have to take a ticket by choosing his own number.

What if two people are choosing numbers at the same time, and one guy goes into the critical section while the other is still choosing? The second guy might choose a smaller number and go into the critical section too.

The way to avoid the bug is to have one process X raise a flag while he's choosing a number, and when somebody Y is trying to decide whether he is next in line, Y will stop and wait for X who is choosing to lower the flag. That algorithm turns out to be correct.

Lamport's simple analogy framed the solution, but the details of his approach were intricate. As a result, the algorithm and its proof enjoyed careful scrutiny for years by researchers who mined it for techniques. Lamport himself used it that way.

Just about everything I did for the next few years came as a result of studying the Bakery Algorithm. I've invented many algorithms, but the Bakery Algorithm is the only one I feel I discovered.

Lamport's Bakery Algorithm has been used in several products, but its principal contribution has been to the general problem of designing systems that would behave correctly even in the face of failures.

I think the property I was looking for was failure-tolerance. Other algorithms [including semaphore-based ones] used a single memory location that was written by all processes. If that location failed, then everything failed.

With the Bakery Algorithm, each memory location is written by only one process, which we consider the "owner" of the location. If a process fails, it causes the failure of only the memory locations that it owns and those locations assume a default failed value. Thus failure of some processes does not prevent the other processes from continuing.

The other important property of the algorithm was that it did away with an assumption even more basic than semaphores. This came as a pleasant surprise to Lamport.

It was only after I wrote out a correctness proof for the Bakery Algorithm that I realized that its correctness didn't depend on what value was returned by a read that was concurrent with a write. This meant that [hardware-enforced] mutual exclusion of access to shared variables was not required. What I had done was discover a mutual exclusion algorithm that didn't assume any underlying mutual exclusion.

Lamport's contribution to theory was solid. It did not depend on his choice of language or on any unrealizable assumptions. It depended only on the laws of nature and the algorithm. That gave him the confidence to extend the result to a technical problem known as the atomic register problem (see Box, "Lamport on Atomic Registers").

My motivation was physical reality. It was almost as though I was thinking more of the physics of concurrency than the computer science of it. That was just totally alien to people.

LAMPORT ON ATOMIC REGISTERS

The Bakery Algorithm led me to ask the question: What was the simplest, most fundamental method by which processes could communicate with one another. I decided, based on physical considerations, that it consisted of a bit that could be written by one process and read by another, with no assumptions about what happened when a read and write happened concurrently—except that the read would get a 0 or 1.

I then showed that you could implement an "atomic register"—one that behaved like the memory assumed in conventional algorithms. I already knew how to use such a register to solve the classical synchronization problems, so this showed that those problems could be solved using only these primitive bits.

I couldn't interest anyone in it, so I never wrote up my results. Around 1985, I saw a paper inspired by problems in integrated circuits which was heading in the direction of my work on atomic registers. Since my assumptions were based on physical considerations, I wasn't surprised that there might have some application in VLSI circuits. So I wrote it up and published it.

This started a small industry, with theoreticians extending the results to implement multireader and multiwriter registers—a problem that I had spent some time thinking about, without success.

Time and Distributed Systems: Lamport Clocks

Having found that physics offered a fresh perspective to fundamental issues of distributed computing, Lamport next drew on his fascination with special relativity in order to study the concept of time in distributed systems.

A *distributed system* consists of many elements that are separated in space and connected by networks that take time to transmit their messages. This is true for an office network, an international telecommunications grid and even the postal system.

But how is time determined in these systems? Let's look at a nationwide computer network. If each computer has its own clock, the various clocks might not be synchronized. (Remember that a few nanoseconds—billionths of a second—can make a difference in a computer system.) If each computer consults a single clock in, say, Dayton, Ohio, then there is some delay before each computer gets the time and the delay may vary. Sending a message from San Jose to the clock may take 1.4 milliseconds when the system is idle or 30 milliseconds when the network is congested. So what does the commonsense notion of a global clock time really mean in a distributed system?

Lamport decided that the notion meant nothing—that it should be abandoned. He took his inspiration from special relativity.

> If you had this yardstick zipping by in time-space, how would you measure it? Well, maybe you would put firecrackers at both ends and arrange that they would explode simultaneously, and then you could see where the puffs of smoke were. The notion of "at the same time" is what's relevant. Observers who are moving relative to one another will have different ideas—this notion of "now" is relative.
>
> In distributed systems just as in relativity, different observers may have different ideas about which of two events happened first.

In his paper "Times, Clocks, and the Ordering of Events in Distributed Systems," Lamport explained how to handle a world without a global clock. According to Lamport, the key question was how to identify pairs of events that must occur in a particular order for things to make sense.

Drawing on the relativity analogy, he asserted that there were three orderings determined by the physics of distributed systems.

1. If a single processing element completes event A before starting event B, then all processing elements will agree that A occurred before B.
2. If event A sends a message and event B receives the same message, then all processing elements will agree that A occurred before B.
3. If all processing elements agree that A occurred before B and all processing elements agree that B occurred before C, then all will agree that A occurred before C.

If you add irreflexivity (an event does not occur before itself) to the third rule, you get what mathematicians know as a *partial ordering*. This partial ordering corresponds to what you might think of as the path of possible causality: if information may have flowed from A to B, then A occurs before B. For a real-world example of this, see Figure 1, which diagrams a situation in which Bob sends a fax to Alice and Alice sends a fax to Harry, both of which are received.

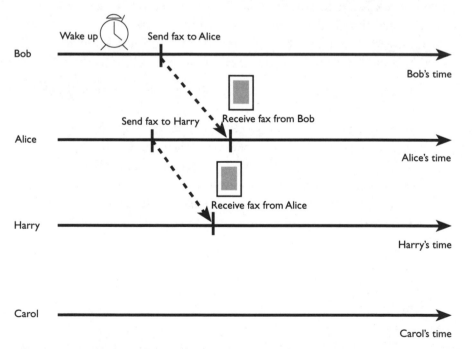

Figure 1   Path of possible causality.

All observers (for example, Carol) will agree that Bob woke up before sending his fax to Alice, that Alice sent her fax to Harry before receiving the fax from Bob, and that each fax was received after it was sent. Notice that some ordering information is not known by any of the participants, for example, whether Alice sent her fax before Bob sent his or even before Bob woke up. Because not all events are ordered, the notion of time is inherently partial. Any algorithm for distributed systems must take this into account.

Lamport's paper then showed that a global ordering could be constructed by consistently extending this partial order. That is, every event is assigned a "time stamp" such that every event has a different time from every other event. The global ordering is now known as a *Lamport clock*.

> The clocks are a collection of clocks, one per process, synchronized according to the following rules:
> (a) Two different events at the same process happen at different times on the process's clock.
> (b) If A sends a message to B, then time on A's clock when the message is sent is less than the time on B's clock when the message is received.
>
> I realized that in order to build a distributed system, the different processes needed coherent views of when things happened. In particular, they had to be made to agree on the order in which any two events occurred. If the events were causally related, then they had to agree on the causal order. If the events were not causally related, it didn't matter which order they picked—but they all had to pick the same order. If such agreement can be reached, then it becomes easy (in principle) to get a distributed system to do anything you want it to.

The universal adoption of ordering according to Lamport clocks does more than throw out the notion of simultaneous events: a Lamport clock ordering may even disagree with the ordering produced by perfectly synchronized (according to a single observer) clocks. That is, event A may occur at 12:01 p.m. on one computer and event B may occur at 12:02 p.m. on another computer as seen by an external observer looking down on the two systems, yet the

Lamport clock may give an earlier time to event B. By their very definition, Lamport clocks ensure that such violations will never matter: since B has a smaller Lamport clock value than A, no information could have flowed from A to B.

Proving the Absence of Failures

In 1977 Lamport left Computer Associates to move to SRI International, a think tank in Palo Alto, California. One of SRI's projects was the NASA-sponsored Software-Implemented Fault Tolerant Multiprocessors (SIFT) project, whose goal was to design fail-safe airplane control systems. The idea was to handle even very rare failures. A fatal error that occurs in one flight out of a million could still cause many airplanes to crash every year. Overcoming such odds requires hardware and software that can "tolerate" failures. Even if a failure should occur, the rest of the computer system must still work correctly. In its simplest form, fault tolerance doubles the hardware. Instead of having a single processor do a task, you have two identical processors running the same program. If one stops working, then the other continues to do the task.

But fault tolerance becomes far more complicated if you admit the possibility that a failing processor may behave unpredictably when the software fails. In the worst case, an unpredictable processor may behave traitorously. Traitors are worse than dead processors, because they might confuse the working processors by sending contradictory messages to them. Benedict Arnold was far worse for the colonists' cause as a live traitor than as a dead patriot.

Before Lamport arrived, the SIFT group (Robert Shostak, Michael Melliar-Smith, and Marshall Pease) had found a way to handle such failures, but at a high price. They had in fact shown that triple redundancy—having three times as many processors as there could be failing processors—was not enough!

Lamport encouraged them to write up their results. Equally important, he gave the problem a name that would suggest intrigue: the Byzantine Generals Problem.

> I decided it wanted some imagery, so I came up with the idea of
> traitorous generals. First, I had the Albanian generals—Albania

was like a black hole in those days, so I figured it was unlikely that any Albanians would object. Then, later, came the Byzantine name.

Lamport's other contribution was to find a new, simpler solution by incorporating digital signatures. Like handwritten signatures, digital ones identify a message from a particular source. (See Figure 2.) Digital signatures are even more secure—no one knows how to forge them except maybe the National Security Agency.

When digital signatures are used, a traitor who sends contradictory messages will be unmasked when other working processors

Figure 2   Simplified example of the problem of traitorous failures. Given the messages he receives, B cannot distinguish between the situation in which C is a traitor (scenario 1) and the one in which A is a traitor (scenario 2). If all messages were signed, however, B would know whether A had modified C's message.

exchange the messages they have received. The working processors will see the inconsistent messages and will know to ignore the faulty processor. Use of digital signatures has become a standard technique for fault tolerance design.

Lamport credits his discovery to being in the right place at the right time.

> This is a nice example of how science depends on informal communication and chance. In those days, digital signatures were a pretty arcane concept. Hardly anyone but computer scientists working on cryptography had heard of them, and there were hardly any computer scientists working on cryptography. I knew about them because Whitfield Diffie was a friend of mine.[2]

Fault tolerance is only one characteristic of safe systems. The other is failure prevention. The SIFT project aimed to eliminate even extremely rare failures in hardware and software design. Since testing cannot cover all possible cases in a complicated program, most rare failures are not identified through testing. This doesn't mean they won't happen in real life.

For example, despite extensive testing, the first space shuttle launch was delayed because of a program error. The test that might have caught this particular error—which would occur only when the computer clocks had certain relative values—had only a 1 in 67 chance of revealing the problem. So the error was "caught" only when the computers failed to start up properly a few hours before the intended launch.

According to Peter Ladkin, who researches reliability at the University of Stirling in Scotland, this problem led to further experiments, some of them quite unnerving.

> After the aborted launch of the space shuttle, the crew practiced a flight-abort procedure in the simulator, which involved dumping fuel and landing in Spain. In the middle of the procedure, the flight

---

[2]Whitfield Diffie and Marty Hellman invented many of the techniques of public-key cryptography and digital signatures.

control processors halted. The problem was found to be a computed go-to (a transfer of control to a location computed by the program). At some point in the computation, the go-to would transfer control to instructions that were not in the computer memory, resulting in paralysis. Programmers found this problem and 17 others, some of which could have serious effects. That was in 1981. Since then, the code size and complexity of digital flight control systems have increased enormously, making such errors even more likely.

The 550,000 instructions in the shuttle flight software rank among the best in the world. Software has never caused a real accident in a shuttle mission (as opposed to an accident that occurs during simulation), nor has it compromised the goals of any mission. Bob Hinson, who is chief of shuttle data systems, says the main testing methods so far consist of design walkthroughs (several people look through a program to determine how it will work) and extremely careful testing. Hinson estimates that fewer than 1 error in 5000 lines of code or about 100 errors in all remain in the flight software. This is a much smaller error rate than civilian industry achieves. Industry doesn't do as well partly because of cost. Once design, coding, and testing have been taken into account, changing a line of code costs roughly $2000.

To Verify or Not to Verify

Lamport and colleagues like Ladkin believe that testing alone will never be sufficient. They believe that one should attempt to *prove*, that is, formally verify, that the hardware and software, particularly the software, are correct. Lamport is one of the strongest advocates of formal verification in the computer science community.

> The verification of SIFT came up with a bug that would not have been discovered any way other than through formal verification. It was the kind of bug that might have caused one airplane to crash every couple of years due to chance cosmic rays.
>
> Verification has to be an important component in the construction of life-critical computer systems, and currently it doesn't seem to be. People depend on testing. You certainly can't eliminate testing, but testing is no substitute for formal mathematical verification.

> Programs and algorithms are abstract mathematical entities.
> We can apply mathematics to them. So, for God's sake, let's do it.

Lamport's view is controversial. Most practitioners avoid verification, probably because it is so difficult. This is also the standpoint of the research community.

> In 1979, Richard De Milo, Richard Lipton and Alan Perlis published a paper "Social Processes in Proofs, Theorems, and Programs," which was essentially very down on program verification. They made some good points. They showed that mathematical proofs by mathematicians are not at first very reliable. Reliability comes to a mathematical proof through a social process in which many mathematicians review an important proof before the community accepts it as correct. That social process is absent in program verification.
>
> Then they introduced a couple of red herrings and came to the conclusion that program verification is impractical. What they actually demonstrated is that the only kind of verification that will work for programs is mechanical verification. People are not going to check proofs—machines will have to. I think that's true of life-critical systems.

This question—to verify or not to verify—is not merely an academic one. Increasingly complex computer systems are put in place every day and society relies on their correct operation to an ever greater extent. Peter Ladkin offers the example of a modern airplane.

> In September 1993, a Lufthansa Airbus A320 crashed at Warsaw airport. The logic of the braking mechanisms (controlling application of wheel brakes, spoilers and thrust reversers) delayed deployment of the braking systems for up to 9 seconds after the plane landed in bad weather conditions. The aircraft crashed over an earth bank, killing two and injuring many. According to the design logic, the airplane was still "flying" during these seconds, even though the pilots had put it on the runway, and were helplessly waiting for the braking systems to deploy.

Lamport has reacted to the problem of hidden bugs by developing new methods for writing hierarchical structured proofs.

Here's how Lamport's method works. At the highest level, the theorem writer makes a basic statement of the theorem. At the next lower level, he or she presents the sequence of assertions that lead to the theorem statement. At the next lower level, the sequence of sub-assertions leading to each top-level assertion. And so it goes—for as many levels as it seems reasonable. This method should make proof checking less painful than it has been.

Lamport's lifelong obsession with seeing proofs done right may well end up preventing planes from crashing and power plants from blowing up. For the moment, Lamport himself sees the hierarchical proof method as only a partial solution.

> My method of writing structured proofs is a way of making hand proofs more reliable. They will never be as reliable as machine-checked proofs, which means that they are probably not suitable for life-critical systems. But they are a lot more reliable than unstructured hand proofs. I think they can provide an economical alternative to the more time-consuming machine-checked proofs for non-life-critical systems or components.
>
> For example, the correctness proof of the part of the system that actually flies the plane might be machine-checked, while the part that calculates the most fuel-efficient flight profile might be verified by hand.

Sidelights

Lamport still remembers the fun side of computing. In fact, he is probably best known for a hobby of his—a formatting system known as LATEX.

> Since the early 80s, I've been writing a book on concurrency. When Knuth's "new version" of TEX came out around 1982, I knew I'd have to write a new set of macros for my book. I decided that I might as well put in a little extra effort and make them usable by others.
>
> The result was LATEX—a multiple pun, "La" as in "the" and in "Lamport," and LATEX as in rubber. Needless to say, a "little extra effort" turned into a lot of extra effort; I think I spent somewhere around ten months on it.

It has become the lingua franca of document-preparation languages for the sciences—I would guess that close to 75% of all computer science papers are written in LATEX. I'm much better known as the author of LATEX than for my real work. People are surprised that I'm not devoting my life to LATEX.

Redefining time for distributed systems, taming traitorous processes, verifying safety-critical code, creating document-preparation languages—Lamport has played a pivotal role in all these fields. Why him and why these problems? His explanation is modest, but comprehensive.

When I look back on my work, most of it seems like dumb luck— I happened to be looking at the right problem, at the right time, having the right background.

Stephen Cook and Leonid Levin
A GOOD SOLUTION IS HARD TO FIND

The idea that there won't be an algorithm to solve it—this is something fundamental that won't ever change—that idea appeals to me.

—STEPHEN COOK

Sometimes it is good that some things are impossible. I am happy there are many things that nobody can do to me.

—LEONID LEVIN

An average 15-year old can understand a proof of the Pythagorean theorem, but the Greek geometers burnt offerings to the gods when they discovered it. Millions of people can whistle parts of Beethoven's *Ninth Symphony,* but few can aspire to his musical genius. There is a basic asymmetry in every good idea—it is easier to recognize than to discover. In computational theory, this asymmetry is a central fact of life, or so it appears.

Consider the well-known traveling salesman problem. Suppose our salesman, Willy Loman, has a sales route, a travel budget, and a booklet of airline prices. He needs to travel from Boston to the nine other cities shown in Figure 1 and back to Boston, staying within his

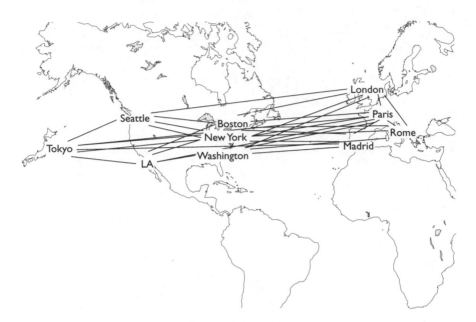

Figure 1   The traveling salesman problem. The task is to determine whether Willy the salesman can travel from Boston to all the other cities and return to Boston within a certain budget.

budget. He asks you to figure out the route he should take. If his budget is large, then your task is easy. If his budget is small, you may have to work very hard. It might not even be possible to stay within his budget. Either way, you may have to consider every possible ordering of cities, and for even this small number of cities, there are approximately 100,000 possible routes!

If there are 3 cities A, B, and C having direct flights between every pair, then there are 6 possible orderings (or sequences for visiting each city): ABC, ACB, BAC, BCA, CAB, and CBA. If there are 4 cities, then there are 24 orderings: the fourth city D can be in four positions with respect to each of the six possible orderings of A, B, and C. For example, with respect to the ordering BCA, we can have the following four orderings: DBCA, BDCA, BCDA, and BCAD.

The number of orderings is known as the factorial (expressed by the ! symbol). Factorials get very big, very fast: 4 factorial or $4! = 4 \times 3 \times 2 \times 1 = 24$; $5! = 5 \times 4!$ which is $5 \times 24 = 120$. $6! = 720$. And so it goes. Computers are good at handling many possibilities, of course, but the sheer number of possibilities overwhelms any amount of computational resources. As Stephen Cook points out:

> If there are 100 cities you have to evaluate 100 factorial tours. No computer is going to be able to try out 100 factorial tours. It's hard for people to understand that. If you do some simple calculations, you realize that if you had all the electrons in the solar system working on it at frequencies comparable with their spins, it would still take until the sun burnt out to find it. The basic point to get across is that there are things you just can't do in practice.

The asymmetry in the salesman problem is this: If someone gives Willy a particular route, it's easy for him to check whether or not it meets his budget constraints. A good solution for Willy may be hard to find, but it is easy to verify.

In the 1950s and 1960s many problems in design, operations research, and artificial intelligence seemed to have this hard-to-find, easy-to-verify property. Many members of the computer science community suspected that these problems belonged to a common mathematical family.

Two scientists working simultaneously and independently, the American Steve Cook and a Russian, Leonid Levin, described this family in the early 1970s. Now known as *NP-complete* problems (for reasons you'll see below), their numbers are huge and growing. For working computer scientists, proving that a problem belongs to this family is tantamount to demonstrating that finding an exact solution for it is impossible—the customer must be willing to accept an approximate solution.

In the Western computational tradition, the notion of the difficulty of computation has roots in logic and Alan Turing's "uncomputability" results. Michael Rabin first formalized the notion of the inherent difficulty of computable problems in 1959. When he defined NP-complete problems in 1971, Steve Cook was following this tradition.

Starting in the late 1950s, Russia's operations research community had informally characterized certain optimization problems as requiring *perebor,* which in Russian means "brute force," or exhaustive. Russian scientists put into the category of *perebor* search those

problems which required a search of all or nearly all alternatives before one could be sure of finding the best solution. For example, trying all alternatives for even a modest-sized version of problems like
the traveling salesman problem can become infeasible.

The theory behind *perebor* problems, however, was inspired not
by logic, as in the Western tradition, but by Andrei Kolmogorov's
notion of the relationship between the randomness of a sequence of
characters and the difficulty in describing it. This approach drew
upon the American Claude Shannon's information theory of the late
1940s. After that time, communication in mathematical sciences between the Soviet Union and the West was sporadic at best.[1]

In talks given in 1971 at Soviet universities and in a short journal paper in 1973, Leonid Levin showed the relationship between
perebor and Kolmogorov's ideas. One of the results was a characterization of NP-complete problems.

This merely started the discussion. Neither Cook nor Levin nor
any scientist since then has actually proved that problems in this
family are extremely difficult. They merely showed that if any problem in the family is difficult, then they all are. They thus posed the
major open problem in theoretical computer science today: Do the
NP-complete problems require exhaustive search or don't they? To
put it another way: An algorithm that could solve problems like the
traveling salesman problem while exploring only a small fraction of
the possible routes would change computer science history.

Stephen Cook: Logic and the Western Tradition

Stephen Cook was born in Buffalo, New York, in 1939. Cook's father, a chemist for Union Carbide, also taught at the University of
Buffalo. He had always dreamed of living in the country, and when

---

[1]The extent of concurrent invention resulting from this mutual isolation is astound-
ing. This chapter is about the independent discoveries of Levin and Cook, but two
other independent results come to mind immediately. While Rabin worked out the
foundations of complexity in 1959, Gregorii Samuilovich Tseitin was doing similar
work in Russia. In 1969 Gregory Chaitin in New York defined a measure of randomness similar to Kolmogorov's, but a few years later.

Stephen was ten, the family moved to a dairy farm in Clarence, New York. They rented the land to a farmer, but kept a cow (which Steve milked) and some other animals.

> I had a normal interest in chess and so on—nothing special. I did well in math in school but it was just an ordinary rural high school. I certainly never thought about being a mathematician. My mother's uncle Arthur was a mathematics professor in Wichita. That's the only mathematics ancestor that I know of.

Clarence, New York, was also the hometown of Wilson Greatbatch, the inventor of the implantable pacemaker. He inspired Cook's teenage ambition to become an electrical engineer.

> I worked for him in the summer in the attic of his barn where he had a little shop. Transistors were new—this was in the fifties—and he was experimenting with transistor circuits. And I would help him solder up the circuits. I found it quite fascinating.
>
> When I started out at the University of Michigan my major was what they called science engineering , but my interest was electrical engineering. My freshman year [1957] I took a one-credit programming course with Bernard Galler.

Cook and a friend concocted a program to test Goldbach's famous conjecture that every even integer greater than 3 is the sum of two primes. As far as they could compute, the conjecture held. (The conjecture remains open, because it is very hard to prove universal properties about prime numbers; on the other hand, nobody has found a counterexample.)

Cook completed a major in mathematics, but he also learned enough computational theory to know about inherently impossible problems such as Turing's halting problem (discussed in the chapter on Rabin). Cook then entered Harvard's Ph.D. program in mathematics with the intention of studying algebra. But his most influential teacher turned out to be Hao Wang of the applied science division. Trained in mathematical logic and philosophy, Wang worked on automatic theorem proving—the discovery of proofs by the computer itself.

Another influence was complexity theory, which had just been given a mathematical foundation by Michael Rabin's 1959 paper.

Many of complexity theory's seminal figures, including Rabin, Juris Hartmanis, and R. E. Stearns, delivered talks to eager graduate students like Cook.

> It seemed like a very natural and basic question. Obviously there are problems that are solvable in principle by algorithms but not in practice, because the sun burns out before you solve them. So, it's just a very natural question to ask about the inherent difficulty of problems.
>
> Before real computers existed, you couldn't execute algorithms except by hand. The process was so tedious that the question of complexity was less interesting. Now that we had these powerful machines to help us and they seemed like an enormously powerful tool—thousands of operations per second—it was very natural to ask just what sorts of problems could you really solve.

Cook's advisor, Hao Wang, added a logical perspective to these considerations.

> I was very aware of his [Wang's] ideas and his techniques. My result for NP-complete problems is an analog of his. Turing and Wang were talking about the predicate calculus. I was talking about propositional calculus.

Predicate calculus and propositional calculus are two languages used in mathematical logic. Wang studied the complexity of the "satisfiability problem" for the predicate calculus. Cook later became interested in satisfiability for the propositional calculus. To understand the distinction, consider the following examples.

Predicate calculus makes statements about groups of individuals. For example, the assertion "All Olympic athletes are fit" becomes: for all x OlympicAthlete(x) implies fit(x)."

Predicate calculus permits you to substitute particular individuals for x. For example, if you know that Achilles is an Olympic athlete, that is, that "Olympicathlete(Achilles)" is true, then the above formula leads to the conclusion that "fit(Achilles)" is also true.

By contrast, propositional calculus, which Cook studied, is a simpler language that only allows you to make assertions about individuals. For example, "Tweety is a bird." The rules of propositional calculus allow you to infer new propositions from old ones. For example, if the propositional sentences "Either Tweety is a bird or Luke is a gazelle" and "Luke is not a gazelle" are both true, then the rules allow you to infer that "Tweety is a bird" is true.

Satisfiability

An important question for both logical languages is whether there is an assignment of true and false values that makes a given formula true. If there is, then the formula is *satisfiable;* otherwise it is *unsatisfiable*. For example, "x and not y" is satisfiable because it is true whenever x is assigned the value true and y is assigned the value false.

On the other hand, "x and not y and not x" is unsatisfiable because either x or not x must be false under any assignment of truth values, making the whole assertion false.

Alonzo Church and Alan Turing had shown that determining whether certain formulas in predicate calculus were satisfiable was computationally impossible even in infinite time. Cook would show that satisfiability for propositional calculus may require trying a large percentage of the possible truth assignments.

How many possible assignments exist? If there is one propositional variable (such as x), then there are two possible assignments (x is true or x is false). When there are two variables (x and y), then there are four possible assignments: x true, y true; x true, y false; x false, y true; and x false, y false. In general, if there are n variables, then there are 2^n possible assignments. This comes to one thousand for 10 variables, one million for 20 variables, one billion for 30 variables, one trillion for 40 variables. It keeps growing by a factor of 1000 for every 10 additional variables.

The number of possibilities is said to rise exponentially with the number of variables n, because n is in the exponent. If an algorithm tests each possibility separately, then the time to execute the algorithm is also exponential and the algorithm is said to take exponential time. By contrast, in a polynomial-time algorithm, time requirements rise according to an expression of n raised to some fixed power, where n is the size of the problem.

For example, the polynomial time expression n^3 has the value 1000 when n is 10. When $n = 20$, the value is only 8000; it's 27,000 for $n = 30$ and 64,000 for $n = 40$. Compare this with the exponential growth of one trillion possible assignments for 40 variables!

Defining NP-Completeness

After completing his doctoral thesis at Harvard, Cook spent a short period at the University of California at Berkeley before moving to the University of Toronto.

In 1971, his ideas about satisfiability coalesced in a paper for the Third Annual Symposium of the ACM (Association for Computing Machinery) on the Theory of Computing. Cook discussed problems for which a possible solution—a "candidate"—could be checked in polynomial time. Since it is not always possible to decide which candidate is a good one to try, a program may have to guess a solution. For this reason, Cook called these problems nondeterministic polynomial, or NP, problems. The guessing part is nondeterministic and the checking part is polynomial.

The traveling salesman problem and satisfiability both have this property, since you can quickly check whether a candidate travel plan satisfies the salesman's budget or whether a candidate truth assignment yields a true formula. The big question is whether it is possible to find the appropriate candidate in polynomial time.

Cook showed that the satisfiability problem is among the hardest problems in NP. Specifically, he showed that if satisfiability has a polynomial-time algorithm, then so does any problem in NP. That made satisfiability an NP-complete problem. In practice, then, showing that a problem is NP-complete implies that it is hard to solve, though you could recognize a solution fast if you found one. And if by some lucky quirk of fate someone found an efficient algorithm for one NP-complete problem, then that algorithm could be used for all NP-complete problems.

Shortly after Cook's paper appeared, Richard Karp of the University of California, Berkeley, demonstrated that twenty-one other problems are NP-complete, including a problem closely related to the traveling salesman problem. He argued by a method known as

---

USING GENES TO FIND SHORTEST PATHS

Recently, Leonard Adleman of USC discovered a way to solve the traveling salesman problem using strands of DNA. DNA consists of two strands of chemical constituents known as nucleotides. There are four kinds, conventionally labeled A, C, T and G. Certain pairs of these bond together (specifically, A with T and C with G) giving DNA its characteristic double-stranded shape.

Adleman represented each city by a sequence on a single strand. If $X^{in} X^{out}$ is the strand for city X and $Y^{in} Y^{out}$ is the strand for city Y, then the flight from X to Y would be represented by $x^{out} dy^{in}$. Here x^{out} pairs with X^{out} (each A in one is at the same position as a T in the other and similarly for C and G); y^{in} pairs with Y^{in}. The d has a length that is proportional to the cost to go from X to Y.

Mixing the city strands and trip strands together, he used well-established biological lab techniques to find the shortest double strands that included all cities. This is the path the salesman should take.

This remarkably clever approach allows one to think of extremely small, low-power computing devices. Does this approach offer a general solution to all NP-completeness problems? Unfortunately not. Solving the traveling salesman problem for 1000 cities would require far more molecules than there are atoms in the known universe.

---

reducibility: if any of these problems could be solved fast, then so could satisfiability. Therefore, these problems were at least as hard as satisfiability—and vice versa. Unfortunately for science, none of this proves that these problems are in fact difficult.

A (Nearly) Perpetual Discussion Machine

So, you might ask, if no problem was really proved to be hard, what was accomplished? An analogy may help. The patent office will immediately reject a patent application if it promises perpetual motion. The patent examiners use a reducibility argument: if the machine worked as promised, then the hypothesis of energy conservation would be false. The examiners have a lot of faith in that hypothesis.

The perpetual motion machine is analogous to an NP-complete problem. If any such problem could be solved fast (in polynomial time), then all could, so the working hypothesis that NP-complete problems require exhaustive search would be false. Computer scientists have a lot of faith in the hypothesis that NP-complete problems are really difficult. Therefore if a computer scientist encounters a problem for which he can't give a fast algorithm, he may try to show that the problem is NP-complete, implicitly invoking the hypothesis that such problems really are difficult. Cook sums up that position.

> NP-Complete problems probably do not have polynomial-time solutions. We know that all NP problems have exponential-time solutions. They may also have polynomial-time solutions, but that's the big open question. If one NP-complete problem such as satisfiability has a polynomial-time solution, then they all do. They're all interreducible—that's the importance of the question.

Since Karp's work, researchers around the world have shown thousands of problems to be NP-complete. Typical examples are the optimal geometric layout of telephone networks, or the best way to play a game like checkers. Cook was taken aback by the number of NP-complete problems.

> I thought NP-completeness was an interesting idea—I didn't quite realize its potential impact.

Leonid Levin: The Kolmogorov Tradition

While Cook was contemplating complexity theory at Harvard in the 1960s, a young high school student was learning about *perebor* and Kolmogorov complexity in the Soviet Union.

Leonid Levin was born in 1948 in Dnepropetrovsk, an industrial city in the heartland of Ukraine. His father, Anatoly Levin, first taught Russian language and literature in high school, then completed a Ph.D in education to teach at the university level. Leonid's mother, Anna Erenburg, was an industrial architect who designed bridges. Early on, Levin became interested in science and mathematics.

> I was frustrated when I learned Mendeleyev's table [the periodic table of elements] It was not quite as regular as I would like it to be. I would work hard and rewrite and rewrite the table. I mostly would put it closer to how I later saw it in America.

John Backus helping his daughter compile a sand castle in 1961.

John Backus thinking early thoughts about form and functional programming in 1969.

A young John McCarthy around the time of his invention of LISP.

Kay's original Dynabook drawing.

A more recent photograph of John McCarthy. (Courtesy Department of Computer Science, Stanford University.)

Alan Kay around the time of inventing Smalltalk.

Edsger Dijkstra in 1955,
around the time of his
shortest path algorithm.

Edsger Dijkstra in 1990.

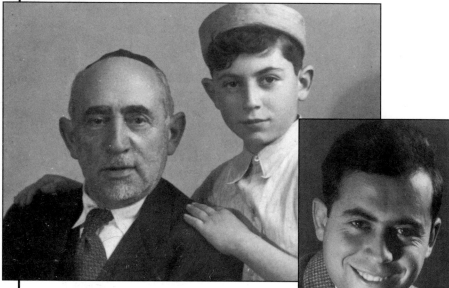

Michael Rabin, age 12, with his father.

Rabin at 22, after getting his
MS from the Hebrew
University of Jerusalem,
about to enter Princeton's
math department as a Ph.D.
student.

Donald Knuth (seated) with Case Western's basketball coach Phillip (Nip) Heim at the console of the IBM 650 in 1959. (Courtesy Case Alumnus magazine.)

Admiral Grace Hopper (designer of COBOL language) gives Knuth the first ACM Hopper Award, 1971.

Donald Knuth and his wife in front of the pipe organ he designed for their Stanford home in 1975. (Courtesy News and Publication Service, Stanford University.)

A young Bob Tarjan in southern California.

Bob Tarjan on a hike through a nonplanar landscape.

An allegorical Tarjan-style self-adjusting tree. (From J. Stolphi, 1987. Communications of the ACM, Vol. 30, No. 3, March 1987, p. 204. © 1987. Reprinted by permission.)

Leslie Lamport, from a recent ID picture.

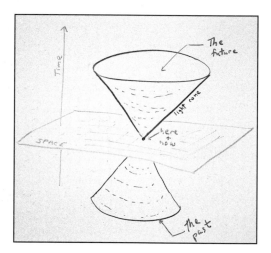

The relativistic light cone drawing that illustrates Lamport clocks. (Courtesy Leslie Lamport.)

Leonid Levin, after a perebor (brute force) haircut in 1967.

Steve Cook sitting at a local maximum in the Adirondacks.

Cook making a conjecture about P vs. NP on the white board.

Levin, during one of his yearly visits to Jerusalem, in 1993.

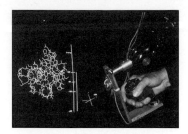

Close-up of a user walking in a kitchen environment. The user is carrying a virtual frying pan to an unshown sink. The head set, designed and built at the University of North Carolina, communicates with an optical ceiling tracker. (Photo by Robert Campell.)

A display combining visual, haptic (touch) and kinesthetic aspects of a computer model of a molecular structure. As the user moves an atom farther from its minimum energy position, the robotic arm pulls back with ever increasing force. (Photo by Bo Strain.)

Fred Brooks in his office at the University of North Carolina, at Chapel Hill. (Photo by Jerry Markatos.)

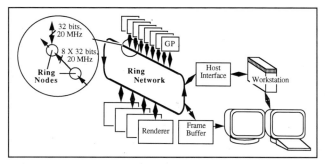

A diagram of the parallel virtual reality computer, the Pixel Planes 5, designed and built at UNC. Pixels are triangular "picture elements" having nearly uniform color and intensity. The more pixels there are, the better the resolution. This machine spreads the work of generating the pixels making up a single image among the nearly one million GPs (general purpose processors) by communicating over the ring network. Each successive generation of virtual reality machines draws more pixels in a shorter time.

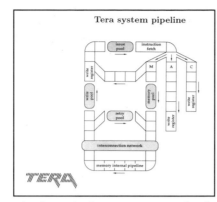

The flow of information in a node of the Tera computer. (Courtesy Burton Smith.)

Danny Hillis holding an inspiring object in his office at Thinking Machines. Some of his other toys are in the background. (Photo by Edward Shashoua.)

Burton Smith serenading his friend Marianne on her birthday.

Hillis in front of a slate thinking machine. (Photo by Edward Shashoua.)

Edward Feigenbaum in his office at Stanford in front of a modern-day Monroe calculator. (Courtesy Department of Computer Science, Stanford University.)

Doug Lenat with daughter Nicole.

Lenat strolling in Antarctica, 1992.

Levin also experimented with chemistry by mixing concoctions in the bathroom of the family's one- bedroom apartment. His father bought him a series of books by the popular Russian science writer Yakov Perelman and also encouraged him to participate in the Olympiads. Olympiads were competitions designed by the Soviet government to locate students with special talent in math and science. Winners advanced from local to regional levels and then were selected for boarding schools specializing in the sciences.

> In Russia in general in the late fifties and early sixties there was a great popular enthusiasm about science. The Olympiads were very popular. Most kids from around the country participated. And then if you won and you were not Jewish, you could go to the international Olympiad as well.

Levin placed first in the Kiev City Olympiad in physics and was sent to the boarding school for physics and math at Kiev University. One day, with all the pomp and ceremony accorded to Soviet scientific "royalty," Kolmogorov himself visited the school. Levin was 15.

> It was a great event when Kolmogorov visited. He met with all the bosses and then he met with the kids. It was in a huge room. He would talk and give a problem and we were told to raise our hands and quite quickly it turned out to be a competition between me and one other boy.

The problems Kolmogorov posed to the young Levin were good preparation for those he would later encounter in computer science. For an example of one of Levin's solutions, see the box on p. 150.
Kolmogorov didn't forget Levin's ability to answer the puzzles and later invited Levin to his own school at Moscow University.

> He taught a small group of two dozen kids. He would make a musical concert for us—he liked music very much. He analyzed poetry. He knew very many things and lots of them on a professional level.
>
> Kolmogorov started his career as a historian and I think he mentioned somewhere that he proved certain conjectures and then he presented his work and they said it's excellent but we need more proofs of that. Then he decided to become a mathematician where one proof was enough.

LEVIN'S SOLUTION TO KOLMOGOROV'S PUZZLE

THE PROBLEM

Suppose all words, all sequences of characters, are of one of two classes—those that are fit to print (decent) and those that are not (indecent). Now given an infinite sequence of characters, can you always break it into finite sequences so all words except perhaps the first one belong to the same class?

THE SOLUTION

Yes. Call a sequence of letters *prefix-decent* if all its initial segments are decent. Suppose an infinite sequence $A = a^1, a^2, \ldots$ has an upper limit n, so that cutting off a prefix of more than n letters always leaves a non-prefix-decent infinite suffix of A (i.e. at least one prefix of that suffix is indecent). Then cut off the first segment with $n + 1$ letters. The remaining (not prefix-decent) sequence has an indecent prefix: cut it off. Repeat the previous sentence forever and you have broken A into finite prefixes all of which (except, maybe, the first one) are indecent. To see that you can do this infinite repetition, suppose you could not. Then after the kth indecent prefix, any subsequent prefix of the remaining sequence would have to be decent, so you could then cut a prefix until that point and get (contrary to the assumption) a prefix-decent sequence.

 If no such n exists, cut off a prefix leaving a prefix-decent infinite suffix. You can repeat the previous sentence forever, otherwise the combined length of the first several prefixes would be the limit n which we assume does not exist. This does not mean that subsequent cuts can be arbitrary, because you might leave a prefix-indecent suffix in that case. But it must be possible to cut it off so that there is a prefix-decent suffix each time (because no such n exists). Now A is broken into finite prefixes all of which (except, maybe, the first) are decent, since they are prefixes of prefix-decent sequences.

Levin immersed himself in mathematics to the exclusion of nearly everything else. He played chess, but refused to study piano.

> I did like to read Russian fiction. I had a very good memory and I would memorize huge pieces of poetry by heart just for sport. Later on, I lost this memory because I was afraid it would limit my imagination.

Although the Soviet system nurtured young scientists in mathematics and physics, it had a history of hostility toward computer science.

> In the early fifties, computer science was kind of an illegal topic in the U.S.S.R. where everything was supposed to be consistent with the teachings of Marx (often interpreted with even greater stupidity than is native to them).
>
> I think some of the Soviet philosophers were irritated by Norbert Wiener—he was a great mathematician when he was young but in his old age he wrote a lot of nonsense about computer science. One of the "pop-science" ideas of his late period was cybernetics, a buzzword which got to denote computer science in Russia. Even though this was kind of harmless nonsense, the Soviet philosophers somehow found it at odds with Marxism.

By the 1960s, however, the Soviet military insisted that computer science be taught because of its evident applications. They even began to train young students in the use of computers, and Levin's first exposure to computers came when he served in the quasi-military units that all undergraduates had to join.

> We . . . worked with some very ancient computers, [using] punch tapes, lots of lights, and panels. It was very impressive. In addition, we learned some kind of probabilistic things. Suppose a rocket is flying. What is the probability it would hit this or that?
>
> I hated the military of course. But the mathematics we were studying were more interesting than our regular university computer course on Algol—just the language nothing else.

Kolmogorov Complexity

For his 1971 doctoral thesis, Levin wrote about Kolmogorov complexity. Kolmogorov served as his advisor and approved the thesis as did Levin's review committee, which included other great Russian mathematicians. Nevertheless, Levin was denied his Ph.D.

> At that time, almost all Soviet youth belonged to Comsomol—an organization like the Communist party but for kids. It was teaching us, through various activities, to love our government. I would sabotage some of this.

For the things I did, there were punishments but it would be bad for them as well: it would show the higher-ups that there was something wrong. So they would try to pretend that nothing happened and I could get away with this.

But I didn't notice that the times had changed after Czechoslovakia—somehow I didn't want to recognize this.

Noisy and arrogant, I was an excellent scapegoat which the university Communist authorities needed at the time.

The biggest blow was not the denial [of the Ph.D.] itself but the quite rare use of explicit political words in the formal justification of the decision. This phrasing prevented me from trying again to get my Ph.D. and eventually led to my emigration.

Before he emigrated to the United States, Levin published a paper in 1973 on Universal Sequential Search Problems in a Soviet journal, *Problemy Perdachi Informatsii*. He outlined a formal relationship between *perebor* problems and Kolmogorov information theory. Kolmogorov information theory explored the relationship between the "randomness" of a sequence of characters (say 0s and 1s) and the length of its description. For example, it is easy to describe a string of a million consecutive 1s in a few words. (We just did so.) Similarly, one can describe any repetitive pattern in a short way, e.g., 00111 repeated a million times.

In Kolmogorov's terminology, long patterns with short descriptions are not random. A fundamental conclusion from Kolmogorov's work was that the amount of randomness depends little on the mathematical language used. For any definition, most sequences are "random"—they require a description at least as long as the sequence itself.

(You can prove this for yourself by a counting argument. How many numbers are there consisting of n binary digits? 2^n. How many different numbers can be described by a program that can be encoded in $m < n$ binary digits? Only 2^m. If $m = n - 10$, for example, then 2^m is only $1/1000$ as large as 2^n, so 99.9 percent of all numbers of length n require more than $n - 10$ binary digits to describe them.)

Levin looked at the problem of how difficult it would be to find programs that could describe long strings.

This amounted to the inherent difficulty of the problem of finding short, fast programs producing a given string. I got an idea to re-

duce tiling [a problem of spatial packing] to it but failed. I succeeded, though, in proving the universality of a similar-looking
problem: finding small depth-2 circuits for partial boolean functions [a problem in circuit design]. I also did it for five other problems including satisfiability, set-cover, graph onto mapping, and
embedding.

Isolated in the Soviet Union, Levin was unaware that Cook and
Karp had independently shown most of the same problems to be
"universal"—or, as they put it, NP-complete.

The Cook and Karp papers were not known in Russia for several
years since the proceedings they appeared in were not received by
any Russian library or institution. Restrictions on travel and cur-
rency prevented this work from reaching Russia by private chan-
nels. Later, I was indeed surprised by Karp's work since I did not
expect so many wonderful problems were NP-complete.

Compared with the reaction to Cook's work, the reaction to
Levin's paper in the Soviet Union was positive but restrained.

Some people were interested. I didn't know if this was a genuine interest or sympathy to my political troubles: people would sometimes be more gentle to me than I deserved. There was no publicity. Then in the late 1970s in Russia people began to get excited
after hearing about Karp's work.

Cook and Levin—After NP-Complete

In the mid-1970s, while Levin was sorting out his problems with the
Soviet authorities, Cook began looking at a new problem: What is
the trade-off between time and memory?

Anybody with a cluttered office or room knows how hard it can
be to find things or to organize a collection of papers in a small
space. With A. Borodin of the University of Toronto, Cook studied
the analogous problem in computers.

Think of the Internal Revenue Service trying to sort a file of 100 million tax returns. Fast methods are known which require a lot of computer memory and slow methods are known which require little

computer memory. We proved that no method can be simultaneously time- and memory-efficient.

[For mathematically adept readers: the product of the time it takes and the memory space used must be at least proportional to $(n^2)/(\log n)$, where n is the number of returns to be sorted.]

To the lay mind, if a scientist states that a problem is impossible or infeasible to solve, that's tantamount to a hopeless situation. Cook reports that his mentor from adolescence, the inventor Wilson Greatbatch, a devout Presybyterian, often teased him about his chosen field.

> Since I got my reputation proving impossibility, he was always very skeptical. He would say, "Making any progress proving things impossible?" [But] it's not that the problem you're trying to solve is impossible. There are always ways around it. You settle for less. You don't really want a perpetual motion machine. You're willing to cheat; you have to bring in an energy source, that's all.
>
> With NP-complete problems, you're going to use heuristics, shortcuts, and approximations and all kind of ways around it. People have studied those. I think the effect of showing NP-complete is to direct people's energies to solving problems that are going to work. I think it's positive and constructive. The thing to emphasize is that although the technology is changing rapidly, there are underlying principles that remain and there are limits.

Cook has continued his fundamental work in algorithms for propositional logic. His current area of interest is finding short (polynomial-length) proofs for propositional logic formulas.

Levin is now a professor of computer science at Boston University, where he has helped develop a theory of "transparent" proofs in collaboration with Mario Szegedy, Laszlo Babai, and Lance Fortnow of the University of Chicago.

These mathematicians have developed a method for writing proofs with the following strange property: if the proof has an error, then a mathematical test on any tiny portion of the proof will reveal an error a certain proportion of the time. For the sake of discussion, let's say it will reveal an error with probability $1/2$ each time. Suppose now that you have just written a proof in this way and want to test its correctness. If you do 40 tests on different parts and find no error, then the probability you have written an incorrect proof is less than 1 in a trillion.

> It's a little like holographic photography. Every part of the photo-graph contains information about every other part. So does every part of the proof.

The Big Question

Recently, Andrew Wiles of Princeton came up with a promising approach for resolving Fermat's last theorem, a problem that has plagued scholars since the eighteenth century, when the French mathematician scribbled a teaser in the margin of one of his proofs. Could the same thing happen to prove that NP = P? In other words, could a problem whose solution can be *checked* in polynomial time also be *solved* in polynomial time? (See Figure 2.)

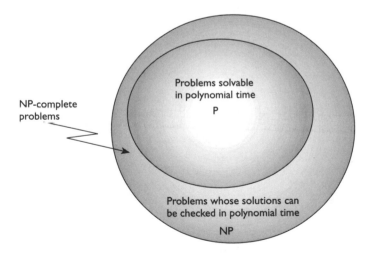

Figure 2   If a problem can be solved in some time, then its solution can be checked in that time. So P is included in NP. The major open theoretical problem in computer science is whether the thousands of important problems that are NP-complete are in fact solvable in polynomial time or do they require exponential time? That is, does P equal NP?

Cook: The sad state of the mathematics of this field is that we can't prove these things. We can't prove P not equal to NP. So it may turn out that someone will invent some clever parallel algorithm that will solve an NP-complete problem in polynomial time.

It's hard to say what progress is in a case like this. It's not just around the corner, but there are partial results; people are attacking it from lots of different areas. And by the way it is possible that P = NP.

Levin: The fact that almost all mathematical conjectures that have been famous conjectures for many centuries have been solved is strong evidence that the solution is polynomial not exponential. Mathematicians often think that historical evidence is that NP is exponential. Historical evidence is quite strongly in the other direction.

ARCHITECTS

HOW TO BUILD BETTER MACHINES

Directing a huge project can be a headache. You must inspire confidence even when you have doubts. You must explore ideas that no one has ever tested. If you are too conservative, the market will sweep you by. If you are too radical, you will risk getting nothing out the door. If it works out, no success is sweeter.

For computer system architects—*system* here means a computer consisting of one or more processors, a network, and the software to run it—even success can have a short life. Other products may incorporate better technology, consigning your organizational ideas to a niche in the folklore of computer construction. Industries depend as much upon folklore as literature does, but authorship is often forgotten when a newer product repackages older ideas. Part 3 of this book presents three computer designers whose ideas are central to the present and future philosophies of computer construction.

In the late 1950s, *Frederick P. Brooks, Jr.* completed a Ph.D. at Harvard and joined IBM at a time when that company's place in the computer industry was by no means assured. His first job was to help design the experimental Stretch computer. Brooks invented the concept of maskable interruptions, a way of controlling the interaction between a processor and its peripherals such as keyboards, touchpads, voice sensors, and hard disks. Today, every computer from palmtops to supercomputers uses Brooks's idea.

In the early 1960s, Brooks led the IBM System/360 hardware and software project which gave birth to a family of machines with interchangeable software that led to IBM's domination of the computer industry for the next 25 years. Brooks left IBM in 1965 to found and direct the University of North Carolina's department of computer science, one of the first in the country. Since then, he and his colleagues have designed much of the scientific underpinnings for the industry now known as virtual reality.

Burton J. Smith, scion of road builders and chemists, was also born in North Carolina, but grew up in New Mexico. An intelligent but bored student in the early 1950s, he built spitball contraptions during science class. At home, he built radios with buttons that would glow in the dark. Halfway through a lackluster college career, he dropped out and joined the U.S. Navy. When he left, he knew he wanted to design electronic devices. After finishing graduate school at M.I.T. in 1972, he briefly joined academia, but soon left to design the first of his many supercomputers. His designs eschew conventional wisdom—processors have no local memories, each processor works on a different task every few billionths of a second, and his networks are simple but very fast. In 1988 he embarked on the quest to build a machine that would do more than a trillion instructions per second. Whether he succeeds or fails, his ideas have already entered the folklore.

As a graduate student at M.I.T., *W. Daniel (Danny) Hillis* pondered a simple question: How is it that our brains, whose neurons fire at best a few thousand times per second, can outperform electronic computers, whose circuits switch in a few trillionths of a second, at so many tasks? He decided the answer was massive parallelism: The brain simply had a lot more "processors" and they work well together to recognize faces and make unlikely analogies. He designed a computer with a million small processors in an effort to build the capability of the brain onto silicon. The pundits of computer science said the idea wouldn't pay off, but risk-takers like CBS president William Paley funded him anyway and Nobel physicist Richard Feynman helped design the first network for Hillis's Connection Machine. Several models later, his machines are used for applications ranging from finding oil to predicting stock prices. Massive parallelism can work, he has shown, provided the problem is big enough and the network is fast enough. Hillis's current research is refocused on biology, as he tries to see whether his machine can reveal the mechanisms of evolution by simulating thousands of generations of life.

Computer architecture is a lot like building architecture. Ideas and motifs arise in controversy, find acceptance, shed their authorship once they are considered obvious, then recombine with newer ideas and technologies. In civil architecture, fundamental new ideas arise every few centuries or so and their origins are lost in mythology. In computer architecture, such ideas arise by the half-decade. Here the idea originators—or at least three of them—tell their tales.

Frederick P. Brooks, Jr.

A DELIGHT IN MAKING THINGS WORK

It is better to take a driving problem that is someone else's problem,
because it keeps you honest.

—FREDERICK P. BROOKS, JR.

cientists are like gods in their research laboratories. They decide the rules of the game—the questions to ask and the methods to use. Their only
obligation is to report their procedures and their results to their
peers. Most scientists, including many great ones, prefer to let others
take results from the idealized laboratory setting to the messy world
outside. Not Fred Brooks.

In 1961, Fred Brooks was on the losing end of a fight to design
and build a specialized line of computers for one division of IBM.
Corporate management wanted a single product line that could
work across all divisions. To his surprise, Brooks was asked to lead
the effort, called IBM System/360. His group succeeded, leading to
IBM's preeminence in computers over the next 30 years.

After leaving IBM, Brooks founded and led the department of
computer science at the University of North Carolina at Chapel Hill.
He has also led a research project in graphics, which has helped lay the
foundations for the industry now known as virtual reality. Brooks's

research philosophy has remained the same in his two careers: Solve someone else's problem and make your solution work for them.

> It keeps you honest when someone can tell you, "What you're doing is just as pretty as you please, but it isn't any help to me at all."

One can hear the gentle cadences of his native North Carolina in everything Brooks says. In fact, he has lived most of his adult life close to his original home. Born in Durham in 1931, Brooks grew up in the nearby small town of Greenville, North Carolina. The son of a medical doctor, Brooks had an early interest in science and computers.

> I got fascinated with computers at the age of 13 [1944] when accounts of the Harvard Mark I appeared in magazines. I read everything that I could get in that field and began collecting old business machines at bankruptcy sales.

The product of a wartime collaboration between IBM and Harvard, the Mark I calculated the trajectories of battleship artillery shells based on such factors as shell weight, angle, wind direction and speed. The electrical counters of this gigantic calculator (50 by 10 by 8 feet) made a great racket when the machine was turned on, but despite its boisterous enthusiasm it was roughly one hundred million times slower than a modern personal computer (it had a 3-Hertz clock). In high school, Brooks caught the postwar fascination with electronics.

> We had a radio club and an electrical engineering club in high school. A very fine high school science teacher, Mr. Robinson, encouraged us in all these extracurricular activities. We did sound effects for plays; we did the music for dances.
>
> In the summers, I banged sheet metal for a living—making flue pipes to go in the barns that cured tobacco. Greenville is right in the center of the tobacco belt.

At Duke University, Brooks majored in physics with a second major in mathematics. His senior project was to build a closed circuit television system out of vacuum tubes. Though he liked physics, he felt drawn to a younger field.

I explained to my physics professor Harold Lewis that my first interest was working with computers. And he said "You can't get in on the ground floor—it's too late—but you can catch the first landing."

The first landing was at Harvard, with Howard Aiken, who had designed the Mark I that Brooks had read about years before. Aiken had parted ways with IBM after the Mark I, because he wanted to use electronic vacuum tubes, whereas IBM continued to prefer electromechanical relays. Aiken made a big impression on the young Brooks.

He was a forceful person. He stood about six foot three with slanted eyebrows and Spockian ears.

When he was mad and glaring down at you, you felt like you might be seeing the devil incarnate. He was usually extremely encouraging. He came to the office every day during the thesis writing period, wanting to read the prose that had been generated since the day before. That's one reason I and some of the other students got through the Ph.D. program in three years.

One of the other students was Ken Iverson, who invented the spartan but powerful programming language APL, for which he won the Turing Award in 1979.[1]

By the time Brooks was ready to begin his thesis work in 1954, Aiken's lab had finished its latest computer, the Mark IV, which was designed for the Air Force. Like its predecessors, the Mark IV was to be used for scientific calculation and engineering.

Aiken was convinced that business computation required a completely different design and asked Brooks to give it a try. Brooks's innovative design led to a job with IBM in 1956 on the extremely ambitious Stretch computer.

Stretch was the world's fastest supercomputer from the first delivery in 1961 until the CDC 6600 came out in 1965. There were nine

---

[1]Iverson and Brooks collaborated on a book, *Automatic Data Processing*, for which Iverson developed APL. At John Wiley and Sons' suggestion, they created two books, with Iverson the sole author of the second, *A Programming Language*. Brooks was acknowledged in the preface of *A Programming Language*.

of them built—it was about a ten million dollar machine. It was designed as a purely scientific supercomputer. The first one went to Los Alamos Scientific Laboratory [the high-energy physics and weapons laboratory].

We built a character processor attachment [scanner] for the National Security Agency called Harvest, which was about three times as large as Stretch itself. A very impressive piece of machinery.

Stretch pioneered many of the concepts of modern computer architecture. First, Gene Amdahl invented and John Cocke and Harwood Kolsky helped develop instruction pipelining.

Pipelining is the computer equivalent of an assembly line. Just as an automobile assembly line moves many different car assemblies at different stages of completion, a pipelined computer moves a collection of operations (adds, subtracts, multiplies, and divides) at different stages of completion. The net effect is that the computer completes a new result in the time it takes to complete a single stage—much less time than to produce the whole result from the beginning.

Cocke would later spur the development of reduced instruction set computers (RISC)—the basis for the PowerPC chip, for example—and many ideas in efficient language translation. Brooks and Dura Sweeney invented the notion of maskable interrupts.

An *interrupt* is like a phone call. Imagine that you are an eager young executive and you have several phones with different phone numbers in your closet-sized, windowless office. To ensure your climb up the corporate ladder, you assign phone numbers to people depending upon their importance to the organization—and your career. Naturally, the president gets the number of phone 1, the vice president gets the number of phone 2, and so on. If the president calls, then you want to ignore or "mask" the other phones from ringing. If you are doing something so important that even the president shouldn't disturb you, then you mask out all phones.

Virtually all computers use this scheme today. Edgar F. Codd (who later won the 1981 Turing Award for inventing relational databases) used the interrupt subsystem on Stretch to build the first interactive operating system. It allowed users to type at a keyboard and see the characters on the screen, a feature we all take for granted now, but which requires the keystrokes to interrupt the computer without causing electronic confusion—it is masking which keeps the computer sane.

The Stretch project left a permanent imprint on the computer lexicon as well. Werner Bucholz, chief architect, coined the word *byte,* meaning a sequence of 1s and 0s long enough to hold a character—he suggested a length of eight. The 256 possible values that can be stored in an 8-bit byte are enough to hold a single character (letter, number, or punctuation symbol) of most European language alphabets. When you ask for a computer with a 900-gigabyte hard disk (in your salesperson's slang, 900 gigs), you are using *byte* in the same sense as the Stretch designers.

As the fifties and the Stretch project wound down, IBM led the scientific computer industry largely because of the success of the lower-cost 7090 hardware and the higher-level programming language, Fortran (see the chapter on Backus). But competition for the 65 percent of the market concerned with commercial data processing was fierce, and RCA, General Electric, Univac, and others looked very strong.

In 1960, after a brief stint at corporate research, Brooks became the head of system architecture in Poughkeepsie, New York, where IBM designed (and still designs) its biggest machines. The system architect decides the general design of a machine—the amount of memory it should have, what types of data it can handle, the details of the instructions it follows, and how it manages its pipeline. Brooks's team proposed a design based on Stretch.

> Top management decided that the corporation did not need a new product line that was localized in one division. There ought to be a new product line across the divisions. My boss got replaced that night; he was succeeded by Bob Evans.

Brooks fought against that decision, but he lost.

> So that fight finally got settled in June 1961. Bob Evans called a retreat of all the managers up at Saratoga Springs from the Poughkeepsie lab to reassign everybody to the whole new set of projects that were the corollary of that decision. I went to make sure my people landed on their feet. I thought I was off to Research again. To my utter amazement they asked me to take charge of the new prod-

uct line—not as architecture manager but as project manager, with engineering, marketing, programming, and architecture under me.

Bob Evans and I had really been at each other's throats during that fight and at one point he called me up and said, "I want you to know you got a raise." I said, "Well, thank you." And he said, "I want you to know that I had nothing to do with it."

Brooks assembled the best team he could. It included not only the division's best architects, such as Gerrit Blaauw, but also Gene Amdahl (who would later start his own company, called Amdahl Computers, in competition with IBM), Jacob Johnson, and Elaine Boehm, all of whom had been at IBM Research building a machine called the ABC. They worked out the high-level design by December. One month later, corporate management approved the development of what was to become the IBM System/360 family of computers.

The concept was simple enough: if a user wrote a program for one machine in the 360 family, the program would run on any larger (and most smaller) machines in the family. This meant that a growing company could purchase a small machine, write software for it, and then use the same software for larger machines as its growth required. (In the boom-boom economy of the sixties, this was an appealing feature.) To achieve such a close family resemblance, each machine had to process the same instructions in exactly the same way and all of the supporting software had to be standardized.

In the mid-1990s, this idea surprises no one. Not only IBM, but Intel, Motorola, and most other computer companies support this idea of standardized instructions. But in the early 1960s, no corporation had taken this position. For example, RCA engineers flatly refused to design a family of interoperable machines. They impatiently explained to their managers that their different machines "couldn't" run the same instruction sets. (By contrast, Gene Amdahl accepted Evan's offer to design the instruction architecture only if he could design one set for all System/360 machines.)

Clearly, in this case, IBM's marketing bias won the day over the technical bias of RCA and its other competitors. The hardware effort went well and IBM was ready to announce the seven machines of the 360 family in early 1964. Or so they thought. The operating system development was hopelessly stalled.

I went to Evans and I said, "Let me go to the other house because that's where the boat's sinking." And he said, "Godspeed," so I became in charge of the operating system for the next year over in the programming systems house.

The first thing I did was to send the team off into the woods to come up with a new plan that became the operating system.

The basic technical innovation in the new design was to exploit the fact that every machine in the family had at least one disk, as opposed to just tapes. As anyone who has ever had to wind a cassette tape to a particular position knows, tapes are *sequential* devices. Moving from point x to point y on the tape requires winding through all intermediate points. Disks, by contrast, are *random access* devices. Moving from point x to point y requires moving a read/write head to the proper track. This is faster and depends little on the distance between x and y. Taking their cue from Codd's work on Stretch, which also had a disk, Brooks and his team realized that the disk permitted them to run programs far larger than the size of the physical memory (which was less than 1 megabyte for the largest machine).

The System/360 operating system first shipped in April of 1965. It worked on seven different computers, from little machines with 32,000 bytes of main memory to machines with 500,000 bytes ($\frac{1}{2}$ megabyte) of main memory. Even though the largest such machines offered less than $\frac{1}{100}$ the speed or memory of one of today's personal computers, companies ran their entire businesses on them. The machines sold so well, in fact, that many IBM salesmen sold their entire year's quota within a few hours after the machines came out.

The hardware design cost for all seven machines was about $200 million. Since every machine had the same core instruction set, the seven machines shared the software cost of $350 million. The designers felt they could celebrate and predicted great things for the future.

We had a 1965 paper that forecast that the architecture was going to last lots of years, while many generations of machines implemented it. And that it would surely go to 32-bit addresses.

The architecture has in fact lasted thirty years. Thirty-one bit addresses (able to fetch data from 2-gigabyte memories) have been the standard since the mid-1980s, and 48-bit address machines are on their way.

As for the System/360 operating system, its direct descendant, called MVS, multitasking virtual storage, still runs on the majority of big IBM mainframes.

In managing that project, Brooks learned some hard lessons about software development, which he set forth in his classic book of essays *The Mythical Man-Month*. Citing other managers in history from Ovid to Truman, Brooks deplored the wasteful practices of corporate software managers, particularly the policy of adding people to a task to speed up its completion. As Brooks wrote: "The bearing of a child takes nine months, no matter how many women are assigned." He also developed a theory of technical project leadership.

> I think it's important to have a system architect who's different from the boss. It's also just as important in the implementation of an architecture to have a chief designer who maintains personal intellectual mastery of the overall design. That's the only way to get lean designs.
>
> In *The Mythical Man-Month* I said build one and throw it away. But that isn't what I say anymore. Now I say, build a minimal thing—get it out in the field and start getting feedback, and then add function to it incrementally. The waterfall model of specify, build, test is just plain wrong for software. The interaction with the user is crucial to developing the specification. You have to develop the specification as you build and test.

The waterfall model holds that abstract specification flows to program construction and on down to test. It remains the method of choice for computer scientists such as Edsger Dijkstra and Leslie Lamport, who propose starting with a provably correct specification and then deriving a correct program from it. For safety-critical software, this makes the most sense—one cannot afford to wait to hear that an airplane crashed before changing the avionics control system. But for research software or even office product software, Brooks and others advocate getting design ideas from end users as soon as possible.

But above all in Brooks's writing in *The Mythical Man-Month* there is a reverence for the creative act of programming.

"As the child delights in his mud pie, so the adult enjoys making things, especially things of his own design. I think this delight must be an image of God's delight in making things, a delight shown in the distinctness and newness of each leaf and each snowflake" (page 7).

In 1964, Brooks decided to leave IBM to establish a computer science department at the University of North Carolina at Chapel Hill. Brooks had profound personal reasons for making the move.

> Fundamentally, I'm a Christian and you go where you're told to go. So I felt that I had a real clear calling to accept this job. The chance to really help people grow is much better at the university level than it is when they're already on the job. And that's part of the calling. Partly it's teaching; partly it's a desire to get closer to the technical work.

At the time, there was only one bona fide computer science department in the country—at Purdue. In those days, the mathematics or electrical engineering departments colonized computer science. Brooks immediately saw the advantages of academic independence.

> We were free to hire and to choose a research direction on computer science criteria instead of choosing on criteria inherited from other disciplines.
>
> Also, we decided that a little department had to adopt what Stanford Engineering School's Dean Frederick Terman [the father of Silicon Valley] called the "peaks of excellence strategy"—pick a small number of fields and try to build a critical mass of faculty in them and let the other subdisciplines go.
>
> I considered software engineering. To my way of thinking, then and now the principal intellectual problems in software engineering are problems of scale, not how to write little programs but how to manage the complexity of big things. And the university is the wrong place to do big software projects.
>
> So we decided that we would pick computer graphics software, and natural language processing.

Entering into his decision was a call to revolution which had just been trumpeted by Ivan Sutherland, the inventor of Sketchpad.

In 1965, Sutherland gave his great vision in his speech at the IFIPS [International Federation of Information Processing Societies] triennial conference, and I was there. The challenge he set forth is one which we're still following today.

He said, "Think of the screen as a window into a virtual world. The task of computer graphics research is to make the picture in the window look real, sound real, interact real, feel real." And I said, "That's where it's at." I've been working on that challenge ever since. Any year we'll get there.

What Sutherland had described is what we now know as virtual reality. But back in 1965, the hardware offered little hope of achieving his dream.

To give the illusion of continuous motion, a computer must display at least 24 frames a second, the speed of a movie. Each frame has to be computed from scratch each time the "viewer" moves in the world.

Brooks's colleagues, Henry Fuchs and John Poulton, now develop state-of-the art machines called Pixel-Planes having hundreds of thousands of processors which draw several million shaded color triangles per second. Even these machines yield only moderate resolution. And they're about 1 million times faster at drawing than machines in 1965 when Sutherland got started.

So full-motion video was out of the question when Brooks launched his department, but he thought his team could work toward that goal on the basis of three criteria.

First, he set out to find a drawing problem appropriate for the technology of the day. Second, he insisted that computer graphics would not only simulate reality but do so with a purpose: to amplify intelligence.

In 1969, I went to the provost, Prof. Charles J. Morrow, and I said, "We are ready and have the necessaries to build an intelligence amplifying system in which the mind and machine will cooperate on tackling hard problems. Who on this faculty most deserves to have his intelligence amplified?"

And then I explained that what I needed was a problem with a lot of geometric content, a lot of computational content, and a lot of pattern recognition.

He said, "Nobody has ever asked me that question before—let me go home and think about it." The next day he responded with a long list of potential collaborators:

Highway safety people on driving simulators, city planners designing low cost housing who wanted real-time, on-line estimating. If you move a wall, how much does that change the project cost? Astronomers concerned with galactic structure, geologists concerned with underground water reservoirs.

In sifting through the choices, Brooks developed the third criteria for his research: not only should computer graphics amplify intelligence but it should also solve problems that affect people's lives. Brooks soon found a problem that matched his research agenda.

One of the folks Morrow mentioned was Jan Hermans, a protein chemist in the department of biochemistry. He had ideas on the protein folding problem.

The still unsolved protein folding problem is to determine the shape of a protein in three-dimensional space given only the sequence of its amino acids.

We started work on that—William V. Wright, later a faculty member at UNC, did his dissertation on implementing a system on an IBM 2250 graphics terminal for visualizing three-dimensional structures of molecules.

Soon after the first project with Hermans, Brooks's group forged an alliance with a crystallographer at Duke, Sun Ho Kim. Kim was trying to calculate the detailed three-dimensional molecular structure of nucleic acid molecules using X-ray diffraction.

The way people did it back then was with brass models and mirrors and plastic sheets with cutout maps of electron density—a thing called a Richards box.

Chemists would build a scale model of a structure they postulated and then adjust it until the model's distribution of electrons agreed with the experiment. One can imagine more frustrating tasks, but not easily.

These things stood about a meter cubed and the bonds were scaled two centimeters to the angstrom. If you had three thousand atoms, it was real hard to get them all right before you knocked some of them wrong. When you were done you would take a plumb bob and a meter stick and measure the 3-D coordinates of each atom.

So we said we ought to be able to build a graphics Richards box which would be much handier. In 1974 that system came up and it helped Kim find the atomic coordinates of the tRNA molecule without building a brass model.

Kim worked as a full partner with the Brooks computer team. His enthusiastic participation confirmed Brooks's notions of having the end user participate in the development process.

Kim would come over and spend eight hours a night moving pieces of his molecule around to try to get a coherent overall pattern that locally satisfied all his electron density measurements.

Later, he stood up and told a National Institutes of Health site visit that he was able to do each week what had previously taken him a calendar quarter in fitting the data.

This successful 1974 experiment also reinforced Brooks's conviction about the relative value of *intelligence amplification,* in contrast to artificial intelligence.

The artificial intelligence approach is to replace the mind. Our approach is always to have the mind at the very center of the system.

Now the artificial intelligence community has come around to this idea after twenty-five years. But that wasn't where they started out. They used to say, "We're going to be able to solve these problems. You don't need a mind." In fact, you do need a mind.

Brooks believes that people will always be better than computers in three kinds of intelligent behavior.

The first is pattern recognition. A month-old baby can identify Mama from directions he never saw her from before and under lighting conditions he never saw before.

The second is judgment evaluation. People walk around a dealer's car lot and choose among the cars using a complex function of many criteria—they do things computers don't do very well.

The third is context searching—bringing to bear apparently unrelated facts. Of course this is a lot of what genius is.

The nicest tale I know in this area was told to me by Tom Watson Jr. [former head of IBM]. He was visiting the underground North American Air Defense command post in Cheyenne Springs, Colorado, and there came an alert. It turned out to be a false alarm. It turned out that their radars had picked up the moon rising.

The commanding general that night was a Canadian. And he said, "Oh, they wouldn't have attacked tonight." And somebody asked him, "Why not, sir?" He said, "Khruschev is in New York."

You could build computer systems with rules elaborated until you were purple, and they wouldn't bring this apparently unrelated fact to bear.

Will computers always fail at these tasks? Each of the five artificial intelligence researchers profiled in this book has a different view of machine limitations. Edward Feigenbaum essentially agrees with Brooks about the need for a symbiotic partnership between human and machine and has accordingly defined the domain of his expert systems. Doug Lenat, in his Cyc project, is trying build into his program enough knowledge about everyday life to overcome the context problem Brooks mentioned. John McCarthy is working on a logical approach to determining how different facts can be related—what Brooks calls judgment evaluation. And Danny Hillis's Connection Machine or its successors may prove to be good at the pattern recognition tasks that humans do effortlessly. But none of these hard-working scientists has yet succeeded.

The research team Brooks leads has constructed other intelligence amplifiers. Using headsets and robot arms, chemists can fly through and manipulate molecules (a technology that the bioengineering company Genentech now uses). Architects can walk through their buildings before they build them, and radiologists can plan the orientation of beams to irradiate a tumor. The team has spurned the entertainment field, however. Many corporate and academic research centers have focused on this potentially lucrative area, which includes what some lascivious hackers have labeled "teledildonics." . . . But not Brooks.

> If what virtual reality becomes is an extension of television, which I consider to be an almost entirely baleful influence on personal development and society, count me out. Somebody else can do that.

If Brooks sounds more moralistic than his peers, the reason may be his strong religious convictions.

> There are certain common ethical standards in science, but Christianity makes you more concerned about carrying them out. It guides you to work on things you think are more important.

Burton J. Smith

RACING WITH
THE SPEED OF LIGHT

Speed is exchangeable for almost anything.
Any computer can emulate any other at some speed.

—BURTON J. SMITH

hen a building withstands the tests of taste and time, much of the credit goes to its architect. The same can be said of a well-designed computer. Like the building's architect, the computer's architect must balance the needs of the user with design and cost considerations. And while modern civil architecture has been a race for height, computer architecture has been a race for speed—faster computers with bigger memories.

The simplest measure of speed is the number of operations executed per second, usually arithmetic operations. The current goal is to exceed a trillion operations per second—about 10,000 times faster than the fastest personal computer.

Since light can travel only one foot in a nanosecond (one billionth of a second) and since electricity in circuits travels even more slowly, there are physical limits to the speed of a single processor. Exactly what those limits are is debatable, but it is an economic fact that each improvement in speed by a factor of 10 is more expensive than the previous such improvement.

When one processor by itself is too slow for an application, the only way to achieve higher speeds is to link many machines together to run *in parallel*. That is, have one processor do a part of the task and at the same time have another processor do a different part. As any student of organizational behavior knows, some tasks cannot be decomposed in this way: the foundation has to be laid before the fiftieth floor can have elevator service.

In organizations, tasks are *parallelizable* if they can be divided into discrete subtasks—each of which can be done without much interaction with others. For example, a bank that is overwhelmed with loan applications can hire more loan officers to process them faster. The same is true in computation. Most scientific computations are parallelizable because their subtasks refer to localized phenomena, e.g., the flow of air across a small portion of an airplane's wing. A programmer can assign a processor to each portion of the problem and then specify how the processors should communicate with one another.

Even when tasks can be parallelized, problems remain; some are technological, but many are logical. Overcoming these problems is the main challenge facing computer architects today. Few individuals possess the required engineering and mathematical skills; and even fewer, the necessary design imagination. Burton J. Smith is one of those few. He has designed two innovative supercomputers: the Heterogeneous Element Processor (HEP) and the Tera machine. (The name *Tera* refers to a teraflop, a trillion floating-point or arithmetical operations per second.) While these machines will not appear in your corner computer store any time soon, the ideas they embody will play a role in high performance computing for years to come.

A Messy Road

Burton Smith was born in 1941 in Chapel Hill, North Carolina. A family history reveals a predisposition to both engineering and science.

Smith's grandfather was a civil engineer who had designed the road leading up to Mount Rushmore. The sculptor of the massive presidential profiles, Gutzon Borglum, was a friend of the family. Burton's uncle was also a civil engineer and his father was a professor of chemistry. In 1945 Burton's father became head of the University of New Mexico's chemistry department, which prompted the family's move to Albuquerque when Burton was four years old. Smith grew up at a time when the radio still held the youth of America in thrall and he was not immune to its charms.

> I was a tinkerer. I built radios, crystal sets with amplifiers. I painted all the light switches in my room with glow-in-the-dark paint so I could find them—and the tools, too. Goodness knows why I was looking for screwdrivers in the dark but they were all phosphorescent.

While Smith devoted hours to radios, academic science left him cold.

> In those days, science was barely taught in school. I can remember in junior high school there were a bunch of us who later on had careers in science who spent the whole class building contraptions to flip spit wads out of rulers. I had a pretty messy road.

Reacting to his lack of interest in school, Smith's parents sent him to private school in California. Smith scored the third highest grade on a statewide high school chemistry exam, but still he was not motivated to study. In 1958 he entered Pomona College and did poorly. In 1959 he transferred to the University of New Mexico. By his second semester at New Mexico, he had grown even more restless than he had been in high school.

> I was studying physics at the time. I discovered I was increasingly less interested in it. I didn't really understand what a physicist's life was like. I've since come to discover that experimental physics might have been very interesting indeed. I wanted something more exciting. That may have been part of my problem—I wasn't turned on by any of the opportunities I saw. I wasn't really ready to launch into a career.

Smith quit school and joined the Navy. For four years, he served on submarines, most often as a communications technician.

I learned a lot and got very excited about going back to school. I decided that designing this stuff would be fun to do.

Smith returned to the University of New Mexico and graduated with a degree in electrical engineering. Smith went on to M.I.T. for a master's degree. Meanwhile, in 1968, he had begun consulting for Hendrix Electronics, a small manufacturer of computer terminals in New Hampshire staffed in large part by M.I.T. graduate students. At Hendrix, his first hardware designs were critiqued by Chuck Seitz, now head of CalTech's computer science department, and one of Smith's few peers as an innovative computer architect.

After finishing his doctorate at M.I.T. in 1972, Smith returned West to teach electrical engineering at the University of Colorado's Denver campus. Soon thereafter, he got work as a consultant for Denelcor Inc., a small company ($1 million annual revenues) with big ideas for general purpose parallel supercomputers.

Assembly Lines, Vectors, and Data Flows

There have been two main approaches to parallel computation: pipeline parallelism (parallelism in time) and multiprocessor parallelism (parallelism in space).

As you may recall from the chapter on Fred Brooks, pipelining is the computer equivalent of an assembly line for mathematical operations. (See Figure 1.) Each circuit element performs part of an operation (such as an addition, multiplication, etc.) in one computer clock cycle and passes its result to the next element, and so on until the operation is complete. At any one time, many operations are in various stages of completion, like partly assembled cars on an automobile assembly line.

For example, in a by-now-dated Cray YMP computer, an addition operation is broken into nine cycles, each 6 nanoseconds long. So it takes 54 nanoseconds for an addition to proceed from beginning to end. With a different addition in progress at each of the nine stages, however, an addition result appears every 6 nanoseconds. This is analogous to an automobile assembly line that outputs a car every minute, even though each car takes 15 hours to produce from

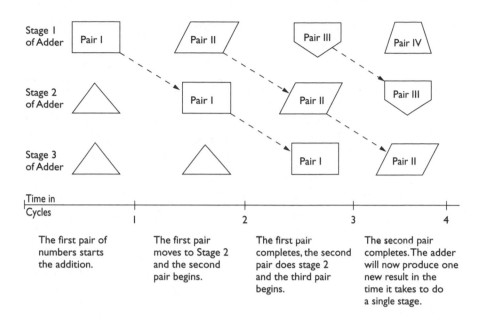

Figure 1   Computer pipelining. Given pairs of numbers, this adder eventually produces a new result in the time it takes to do a single stage of the adder.

beginning to end. Pipeline machines, like assembly lines, work well on problems in which many similar operations must occur in large batches. Pipeline parallelism was the basis for the vector, or array, processors that Seymour Cray, Control Data, and Texas Instruments designed in the 1970s.

Vectors are lists of numbers. Adding two vectors together therefore means adding each pair of elements in two long lists. This is a perfect application for pipelined processors, provided the vectors are long enough. (The batches must be large or the overhead of the multistage pipeline becomes intolerable—that's why vectors aren't used in commercial data processing.)

Vector machines became popular for the vector-rich scientific calculations required for such applications as airflow modeling and weather prediction. The users were bastions of the scientific establishment like the U.S. Naval Research Laboratory, the U.S. Geophysical Fluid Dynamics Laboratory, and Los Alamos National Laboratory. These institutions were unlike the hold-my-hand corporate customers that IBM catered to in the 1960s and 1970s. If something didn't work, the laboratory scientists would figure out how to fix it.

For example, Cray Research lent an early machine to Los Alamos. It had no operating system and no compiler; in other words, no supporting software at all. Los Alamos scientists modified an existing compiler to suit the new machine and loved it. But while vector machines were all the rage at Cray and elsewhere, Denelcor wanted to attack a problem for which the vectors were too short to make pipelining a good strategy: solving nonlinear ordinary differential equations. These equations come up in control applications, often for aerospace problems concerning stability at high speeds. The inherent parallelism of such equations depends on the particular data input, and so it is hard for a programmer to anticipate where parallelism can be used in solving them. Denelcor's dream was to have the machine find the parallelism for the programmer.

Smith and his colleagues at Denelcor wanted to design a machine that would perform an operation as soon as its inputs were ready. This approach is called a *dataflow architecture,* and was first developed in the 1970s at M.I.T. by Jack Dennis and his colleagues. In dataflow architectures, the availability of data rather than a fixed ordering of instructions determines which operation is executed and when.

Dataflow computers resemble an emergency ward. Patients with different symptoms enter at unpredictable moments and doctors perform the appropriate procedures on the spot. In dataflow architectures, data and operation codes enter and one of the processing elements performs its appropriate task. (See Figure 2.)

Different processing elements with different subspecialties, some performing multiplications and divisions, others performing additions and subtractions, execute in parallel. Since each processing element does an entire operation, the result is called "parallelism in space," to contrast it with the "parallelism in time" of pipelines. In 1979, Smith left the University of Colorado to work full-time at Denelcor.

> The thing that caused me to leave the University of Colorado and go to work full-time at Denelcor was an insight I had about mem-

Data   **Operators**

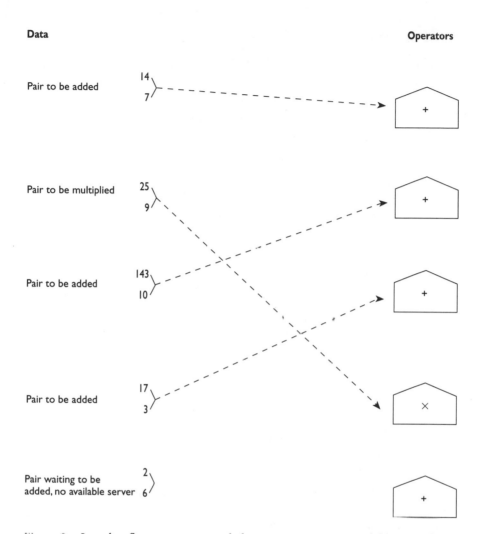

Pair to be added

Pair to be multiplied

Pair to be added

Pair to be added

Pair waiting to be
added, no available server

Figure 2   In a dataflow system, several electronic servers are available to perform operations, here arithmetic ones. When the data elements are ready, they are passed to a server if one is available.

ory. The same machinery that was able to tolerate addition and multiplication latency [the dataflow idea] and asynchrony, could also tolerate memory latency.

I made sure to get tenure first, so they wouldn't say—"we know why you left."

Slow Memories Meet Fast Processors

Memory latency is the time it takes a processor to access its memory chips.[1] The time required, about 100 nanoseconds (100 billionths of a second), may not seem long, but it's 20 to 50 times as long as it takes a processor chip to execute a single instruction. Since many operations require a processor to access its memory chips, speed of processing is not by itself enough to ensure computational speed.

The most common solution to memory latency was and still is the cache. Invented at IBM in the 1960s, a cache is a high-speed but expensive memory that can be accessed in the time it takes to execute one or two instructions. Because of its expense, its size is usually from $1/1000$ to $1/100$ the size of the main memory.

A cache holds copies of the data items that are in main memory. It may seem that such a small memory would be of little help, but a good algorithm for keeping most of the useful data items inside the cache is to give highest priority to data items that have been most recently read or written. When a new data item enters the cache, the algorithm kicks out the item with lowest priority, that is, the one that was least recently read or written.

This simple scheme (known as the *least recently used,* or *LRU,* strategy) works extremely well for uniprocessors. A cache $1/100$ the size of main memory can routinely satisfy 90 percent of the memory requests.

In cache-based multiprocessors, each processor has its own private cache. Therein lies the problem. Since a cache holds a copy of data items that are in main memory, different caches may (and often do) contain copies of the same data item. When one processor changes data item X, all processors must be told to change or discard copies of X, lest they act on an out-of-date value of X.

The same thing happens in everyday life. When many people are writing a document together, each person can hold a copy of the en-

---

[1]Random access memory (RAM) chips are memory devices without moving parts. When your processor looks for data in your laptop, it looks at the RAM chips in main memory before searching your hard disk. A single RAM chip holding 64 million bits of information or more will soon be commonplace.

tire document in a local laptop computer. If one writer modifies a section, then the other writers must be informed of the change so they won't make an inconsistent modification to that section. The amount of communication this requires is large in everyday life, and errors can occur. In the multiprocessor computer case, the circuitry is both expensive and error-prone.

Having decided that the cache solution might not work well, Smith reasoned that the real problem is not that the memory is slow—it is that the processor is not doing useful work while waiting for the memory to return data. So, he designed the HEP to run many tasks on each processor (up to 64).

In one clock cycle, one task would perform one instruction and send requests for the data for its next instruction. At the next clock cycle, another task would do the same thing. Thus by the time the first task issued its next instruction, perhaps 64 cycles later, the data would have probably already returned from memory. Smith is legitimately proud of this idea.

> I saw a way to build machines that can be as big as you please and allow the programmer not to worry about how data are placed in memory. There still remained the problem of how to build the network.

The *network* is the connection between the processors and the memories. Like a network of highways connecting cities, each intersection requires traffic routing logic. The key challenge was to find a good algorithm for routing each message from a processor to a memory. The potential problem was an electronic traffic jam in which the switches become saturated. Traditional networks use a *store-and-forward* scheme in which the switches store messages that they cannot send towards their destination. These message stores are called *queues*. To Smith, queues had problems of their own.

> But what do you do if the queues fill up? It seemed messy to me.

Routing Hot Potatoes

In the spring of 1977, Smith had the idea that if a message arrives at an intersection but can't go toward its destination, then the switch

---

OVERCOMING MEMORY LATENCY

In the cache scheme used by most commercial multiprocessors, one hopes to lay out a computation so that the data for processor A is in cache A, for processor B in cache B, and for processor C in cache C. (See Figure 3.) When this isn't the case, the processor must wait from 100 to 1000 clock ticks. Moreover, the interconnection network must ensure that no two caches are ever inconsistent. Such networks are expensive.

Burton Smith's designs don't use caches. Instead, processors issue requests to remote memories and do other work while they are waiting. Each processor may work on hundreds of tasks at once. The trick is to design processors that can switch from one task to another very fast.

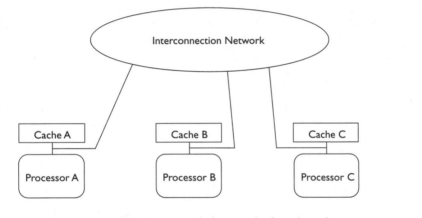

Figure 3   Interconnection network for a cache-based machine.

---

should send the message somewhere—even away from its destination if necessary—just to keep it from causing electronic gridlock.

Paul Baran at the Rand Corporation had come up with a similar scheme for telephones, calling it "hot potato routing." In hot potato routing messages are in continuous movement until they reach their destination. This method eliminates the need for switch queues, but allows misroutings: a message can be diverted from its destination. Smith came up with a more reliable solution.

> I was thinking, that if you keep bouncing them around, you have a battle scar count on each message for the times that it had been

misrouted. When that count gets to its maximum value, then the routing algorithm changes, and instead of getting in conflict with its friends, it goes on an Euler tour of the network.

An *Euler* tour (pronounced "oiler"), named after the great eighteenth-century Swiss mathematician Leonhard Euler, is a route that starts and ends in the same place and goes across every link exactly once. Such a scheme guarantees that every message reaches its destination. Smith likens this strategy to the brownian motion of particles in physics, then justifies the analogy:

> I think about how objects flow, not three-dimensional abstractions.

Even with its 16 processors and clever design, HEP I was not fast enough to find a market—those who wanted high vector performance didn't need a parallel data flow machine; for those who wanted a parallel machine it wasn't fast enough, because its processor components were too slow. Even though the HEP II would improve the speed of its predecessor, marketing HEP I proved to be difficult. By 1985 Denelcor had gone bankrupt.

Although Denelcor was dead, the idea of a general-purpose parallel computer was not. At the Institute of Defense Analysis's Supercomputing Research Center, Smith worked for three years on the Horizon machine, which would be a faster version of the HEP with multiple operations per cycle.

The Tera Machine

In 1988, Smith became, as he recalls, "employee number two" at the Tera Computer Company in Seattle, Washington.

Founded by Jim Rottsolk, the former chief financial officer of Denelcor, the corporate mission was once again to create general purpose supercomputers—capable of performing a trillion operations per second (about the computational capacity of a human brain). While the HEP had linked only 16 processors, the Tera was 16 times more massive: 256 processors arranged in a three-dimensional mesh cube—$16 \times 16 \times 16$. There are 512 memory units each with several memory banks, and 4096 interconnection nodes. The clock period, or the time it takes for an operation, is less than 3 nanoseconds (three billionths of a second).

Having a larger number of processors implies that more messages will be running through the network at the same time, causing more collisions and therefore more unfavorable routings. This in turn implies that the effective memory latency may be even higher than the factor of 50 resulting from the ratio of processor speeds to memory speeds.

One solution is to increase the number of tasks that each processor works on to more than 64; however, decomposing a gigantic job into more than 16,000 tasks (64 per processor times 256 processors) is difficult.

While on a plane to Edmonton, Alberta, Smith sipped a Scotch and began to think about abstract pipelines. He had a revelation.

> Suppose you want to issue eight things, five goats and three sheep [the animals represent requests to memory that don't depend on one another]. I got to thinking that all you really need to do is decide whether the whole group is completed. It clicked on the plane. I jumped out of my seat. I explained it to the woman sitting next to me and her eyes glazed over. Thinking in terms of groups allows the Tera to issue multiple memory references out of one instruction stream—as many as eight.

Smith's key insight was that different operations within a task may sometimes be executed slightly out of order. Memory requests for future operations can therefore take place at the same time as requests for the next one in the task. In contrast, using the HEP strategy, a given task never has more than one request outstanding at a time; if that request is delayed, then so is the task.

Smith was sure that this simple idea would eliminate the need to decompose tasks further.

> I didn't need to test a thing. There were just a few details that needed to be worked out. It's the nature of the idea.

The Attack of the Killer Micros

Part of Smith's quest has been to design a truly general-purpose massively parallel computer. Unlike some pure forms of research in computer science, that quest depends on the vagaries of the marketplace.

As of this writing, the going has been rough. Powerful micro-computers, what Eugene Brooks of Lawrence Livermore Labs calls "killer micros," are making parallel computers seem too expensive. However, Smith remains optimistic, believing that parallel computers will become as commonplace as personal computers are today.[2]

> Ten years from now we will have parallel computers on our desks. In fifteen years we will be carrying one around—it will be much lighter. You'll be able to access parallel computers without worrying about where they are. Maybe we'll redefine general purpose as parallel. Maybe the things that need to be fast will all be parallel.

Parallelism intrigues Smith outside of work as well. A skilled amateur baritone, he sings in choral groups with a special interest in the polyphonic Renaissance music of such sixteenth-century composers as Palestrina, Tompkins, and Mundy.

> It's a little esoteric but it's truly the most rewarding music that I know. It's quite difficult to sing a capella polyphonic music—a lot of melodies going at once. There's nothing quite like it.

A burly man, Smith resembles a former football player. With 40 people to manage, he quarterbacks the Tera team—calling the plays, but letting individual team members find their own way to solutions.

> You can't keep track of everything in a design given the number of people involved. But you have to have some sort of idea of what you want to do. That might be the most important thing: to have the imagination to think of things and get people to do it.

---

[2] Smith's belief is already borne out in one domain: on-line database transaction processing (used in airline reservation systems, banks, and telecommunications). Two vendors, Teradata and Tandem, have built parallel computers for those applications at enormous profit. Even Cray Research builds machines for database processing now.

W. Daniel Hillis

THE BIOLOGICAL
CONNECTION

Clearly, the organizing principle of the brain is parallelism.
It's using massive parallelism. The information is in the connection between
a lot of very simple parallel units working together. So, if we built a computer that
was more along that system of organization, it would likely be able to do the
same kinds of things the brain does.

—W. DANIEL HILLIS

t's hard to imagine how two scientific cultures could be more antagonistic than computer science and biology. Computer science grows out of mathematics, physics, and engineering, where reduction to simple basic principles results in the best designs and most penetrating insights. Modern biology grows out of observing and at times manipulating nature, where complexity presents constant surprises and efforts at reduction usually founder. In their daily work, computer scientists issue commands to meshes of silicon and metal in air-conditioned boxes; biologists feed nutrients to living cells in petri dishes. Computer scientists consider deviations to be errors; biologists consider deviations to be objects of wonder.

W. Daniel (Danny) Hillis practices both disciplines, drawing analogies from each to suggest progress in the other. He has imagined and then built computers modeled on the brain with thousands of processors, even when others predicted the effort would be a waste. His current design calls for more than 16,000 processors with applications ranging from seismic exploration to medical diagnosis. Hillis has turned his Connection Machine computer loose on biological theory and suggested evolutionary mechanisms that only the most daring biologists would consider.

In the Sticks

Hillis was born in 1956 in Baltimore, Maryland. His father, Bill, served as an Air Force epidemiologist, and the family moved frequently on the trail of hepatitis outbreaks.

> I grew up all over the place. I lived in lots of different countries in central Africa—Rwanda, Burundi, Zaire, and Kenya. We were typically out in the middle of the jungle so I was just taught at home.

His mother, Argye, did much of the teaching and was especially interested in mathematics. Hillis's father encouraged his interest in biology.

> When lab equipment was being thrown out, I'd get glassware and use it in my lab. My best biological experiment was tissue culturing a frog heart and keeping the heart beating even while it was growing in the test tube. It was amazing to me that somehow they got together and did this coordinated activity even though they were just this homogenized mass of cells.
>
> I guess early on my exposure was biological. But I would say that my natural inclination was toward engineering. So as a kid I was always building things.

Hillis started with Erector sets and blocks, but soon turned toward engines and robots. A book called *Mike Mulligan's Steamshovel,* with its central image of a little steamshovel building a large courthouse, especially inspired him.

> I always had this picture of how they converted the steamshovel to the heater at the bottom of the courthouse.
>
> I somehow knew about computers too. I had a book called the *How and Why Wonderbook of Robots*—it sort of talked about

computers. It had lots of pictures of things that I now understand were mostly remote manipulators. I had a toy robot too, with a little man inside its head. The little man must have controlled the robot. I remember being fascinated by that concept of the little man inside your head.

At the age of nine, Hillis built his first "computer" out of a record player and two disks. One disk had trivia questions on it. The other had a lot of possible answers. The record player spun around and a coupler caught the two disks and lined the question up with the answer.

While the family was living in Calcutta in the late 1960s, Hillis first became acquainted with the underlying theory of digital computers.

> When I was in India there wasn't any technology around. Even technical books written in English were very hard to come by. The British consul had a library and in the library was George Boole's book called *An Investigation of the Laws of Thought,* in which he lays out what we now call Boolean algebra. I loved the title of the book.
>
> The book was too advanced for me, but I remember getting the basic idea of *and, or,* and *not* out of it.

For a budding computer scientist, this book, first published in 1854, was the right one to read. Boole described an algebra that converts logical propositions to a kind of arithmetic (see box, "Boolean Algebra 101").

Since the fundamental circuits of computers perform and, or, and not operations, Boolean algebra is exactly the right intellectual tool for designing the adders, shifters, and decoders that make up the guts of a machine. That is, if you can get the parts.

> I started thinking about building a computer with the stuff around—all that was around was flashlight technology. So you could get wires and lightbulbs and flashlight batteries, but you couldn't get switches. But I built my own switches using screen doors and nails. So you'd throw a switch by sticking a nail into one screen or sticking it into another screen. The nail had a wire

BOOLEAN ALGEBRA 101

Consider the following three propositions which can each be either true or false.

Proposition R: "It is raining."
Proposition T: "It is Tuesday."
Proposition U: "I am carrying an umbrella."

Now, suppose someone presented you with the following sentence, which combines propositions R, T, and U:

It is not raining or I am not carrying an umbrella and either I am carrying an umbrella and it is Tuesday or it is raining and it is not Tuesday or I am not carrying an umbrella and it is raining.

If you knew that it was Tuesday, raining, and I have no umbrella, you would soon be able to determine whether the sentence was true or false, but it might take some thinking.

Boole's method started by identifying truth with 1 and falsity with 0. He then found operations corresponding to the logical connectives *and, or,* and *not*. Let multiplication have its normal meaning, but restricted to 1 and 0; this will correspond to *and*. Let addition have its normal meaning too, except that $1 + 1 = 1$; this will correspond to *or*. Let the negation sign flip between 1 and 0 so $-1 = 0$ and $-0 = 1$; this will correspond to not.

These definitions allow us to convert the sentence above to the much more succinct formula $(-R + -U) \times ((U \times T) + (R \times -T) + (-U \times R))$. Since we know that it is Tuesday, raining, and I have no umbrella, $T = 1$, $R = 1$, and $U = 0$.

Using these values, $(-R + -U)$ is the same as $(0 + 1) = 1$. $(U \times T) + (R \times -T) + (-U \times R)$ is the same as $(0 \times 1) + (1 \times 0) + (1 \times 1) = 1$. So the whole sentence corresponds to $1 \times 1 = 1$ and is therefore true.

wrapped around it. So I sort of figured out that you could build a computer out of those—a tic-tac-toe playing thing out of screen doors.

The family left India for Towson, Maryland, so Danny could attend junior high school and his mother could begin a doctoral program in biostatistics. Not content with the standard laboratory exercises in biology, Hillis cultured some bacteria found near the gym

and discovered they were toxic. Although the school authorities refused to believe the teenager's findings, fortunately nobody became
infected. By the time he was in high school, Hillis looked elsewhere
for scientific and engineering challenges. The Johns Hopkins chemistry department had lost the manuals to a computer and asked Hillis
to hook the computer up to a mass spectrometer. He programmed it
to calculate and average the results of their experiments. It seemed
easy compared to screen doors and nails.

Hillis entered M.I.T. in 1974 determined to find out how the brain
worked, so he planned to major in neurophysiology. That didn't last
long.

> I ran into Jerry Lettvin the first night I came to M.I.T. He had written
> a paper called "What a Frog's Eye Tells the Frog's Brain"—an idea
> that I got very excited about. He asked me what I was going to study.
> I told him I was going to study neurophysiology. He said "I defy you
> to tell me any good paper that has ever been written on that."
>
> Lettvin convinced me that if I really was interested in how in-
> telligence works, I should try to figure it out by looking at neurons.
> And Lettvin convinced me to go to [Marvin] Minsky.
>
> Of course I had heard of the AI Lab and I had heard of Minsky.
> Minsky seemed kind of an unapproachable character at that point
> so I sort of went and hung out in the lab.

Minsky and John McCarthy had founded M.I.T.'s Artificial Intelligence Laboratory in 1958, and it remains one of the leading AI laboratories in the world. Hillis decided that the best way to approach
Minsky was to help fulfill the laboratory's contractual arrangements
with its funding agencies. Since part of the laboratory's support at the
time came from a National Science Foundation project to use computers in education, Hillis read the proposal to find out what remained
to be done.

> The proposal said one of the things they wanted to explore but
> hadn't figured out was how to make a computer terminal for kids
> that don't know how to read and write.
>
> So I sat around for a while and figured out a way to do that
> and went in and sure enough got a job to do that. I started work-
> ing at the AI Lab in the LOGO group.

Radia Perlman and I developed a terminal that could move icons around. It was before the days of graphical user interfaces, so it was hot stuff.

LOGO was largely the creation of Seymour Papert. Papert had wanted to use computers as an educational tool, so LOGO was a simple programming language—simple enough so that children could generate their own cartoons. (This project had earlier inspired Alan Kay in his design of the Dynabook.) Hillis still hadn't met Minsky, but found out that Minsky was building a computer in the basement.

So Hillis went to the basement, read the plans, and implemented some changes to Minsky's computer. Minsky welcomed the help and took him on as a student, even giving Hillis a room in the basement of his house. Driving to and from work every day gave the two plenty of time to talk.

Minsky's big contribution is the idea that in fact the brain is a complex thing that has multiple goals that are interacting, conflicting, and compromising. So in a sense, what Marvin has done is to take what Freud did much further.

Freud suggested that maybe you have two or three things going on inside of you. Marvin has given us a new image of the mind with a thousand things going on inside of you.

The idea that thousands of interacting decentralized agents create an outwardly unified mind may seem implausible at first, but it follows an evolutionary principle known as *emergence*.

In organisms, genetic mutations seldom result in changed physical characteristics. The physical changes that do arise as the result of mutations are almost always pernicious. Yet, somehow, varied species have evolved through mutation and natural selection to become remarkably beautiful and adaptable.

The principle of emergence states that interacting agents will adapt, through a process of selection, a mechanism for survival.

The concept of emergence appealed to Hillis. He suspected something similar was going on in the brain and resolved to design a computer to test this suspicion. His idea for the Connection Machine was simple. Take thousands of little processors—the original design called for a million—each a computer in its own right with its own memory, interconnect them, and program them to interact and see whether a brainlike intelligence emerges.

> The Connection Machine came from realizing the following con-
> tradiction: the brain manages to outcompute the fastest computers.
> The brain is made out of components that take milliseconds to
> switch whereas the computer is made out of components that take
> nanoseconds to switch.
>
> So clearly the organizing principle of the brain is parallelism.
> It's massive parallelism. The information is in the connections be-
> tween a lot of very simple parallel units working together. If we
> build a computer that was more along that system of organization,
> it would likely be able to do the same kinds of things the brain
> does. The motivation for the Connection Machine was as simple as
> that.

Simple or not, massive parallelism was controversial. The great uniprocessor computer designer, Gene Amdahl (a colleague of Brooks at IBM, but later the founder of a competing company bearing his name) had observed that in many cases there is an inherent limit to parallelism.

An analogy may help to understand his argument. Suppose that a company is willing to hire as many technicians as it takes to pro-duce 10,000 widgets an hour. Each technician will work on one wid-get at a time and each widget takes an hour to build. Before build-ing each widget, the technician goes to the stockroom clerk to get the necessary parts. The clerk requires only 36 seconds to provide parts for one technician.

The company has observed that 10 technicians produce 10 wid-gets an hour. Fifty technicians produce 50 widgets an hour. So the company hires 10,000 technicians, but finds that they produce a measly 100 widgets an hour. The reason: the overworked clerk can provide parts for only 100 technicians in the 3600 seconds that make up an hour.

Amdahl's law encapsulates this observation: it says that if a par-allel task has a sequential portion (getting parts from the clerk in this example) that requires a fraction f of the time ($1/100$ in this example), then even massive parallelism will never speed up production (or computation) by more than a factor of $1/f$ (100 in this example).

Amdahl's law convinced Hillis's contemporaries in the early 1980s that massive parallelism wouldn't work. But Hillis's understanding of the brain made him optimistic.

> There were all of these proofs like Amdahl's law that massively parallel computers couldn't possibly work. And I guess it wasn't so much that I saw what was wrong with those proofs, but I knew that the brain did work and the brain was organized in that way.
>
> So that whether this was a good way to build a computer in general or not, it was a good way to build a computer to do commonsense thinking kinds of things—image recognition, memory retrieval, reasoning—all the things the brain does. That was really the motivating philosophy for the Connection Machine.

In 1983, encouraged by the simulation results of his machine design, the 27 year old Hillis (with Minsky's advice and encouragement) founded Thinking Machines in Cambridge, Massachusetts. Financial support came from both the public and private sectors— CBS president William S. Paley was a major investor, and DARPA agreed to buy the first machine.

A few months before launching the company, Hillis discussed his ideas with Richard Feynman, the Nobel Prize–winning physicist. Feynman pronounced the project "dopey" but then promptly agreed to work for the company the following summer. Demanding "something real to do" when he arrived, he helped analyze the interconnection network of the first machine, determining how much short-term memory each switch in the network would need. Besides his practical contributions to Thinking Machines, Feynman provided Hillis with much needed moral support.

> I was brought up with an academic bias that somehow science is better than engineering. Feynman convinced me that that wasn't true, that there is real value in building things and creating things.

Drawing on his own experiences at Los Alamos during the Manhattan Project, Feynman also suggested how to organize the corporate effort. His main suggestion was that the company pick experts in important aspects of building the machine—software, packaging, and so on. Each expert would become the group leader. (Bill Gates follows a similar philosophy at Microsoft, where he insists that managers have more technical knowledge than their programmers—rarely the case in most high-tech companies.)

Another Feynman innovation was to encourage scientists to offer new applications for the Connection Machine. This led to an early collaboration between Thinking Machines and the neural network pioneer John Hopfield of CalTech.

As defined by Hopfield in 1982, neural networks are a collection of "neurons," each simulated by a computer program. The program for each neuron takes inputs from other neurons or from some outside stimuli. The program then performs some calculations and sends the output to other neurons. This is roughly the way neurons in the brain work.

Hopfield showed how to "train" a neural network with a set of stimuli having known correct responses. The network takes each stimulus and computes its response. If the response is correct, then nothing changes, otherwise the Hopfield network adjusts the calculation at one or more neurons to bring the response closer to the correct one.

Hopfield showed how the network could find patterns in images. There have been many more applications since then, including many in stock market analysis and foreign exchange trading. Neural networks are even known in Hollywood: the character Data of *Star Trek* embodies one.

The Hopfield approach resonated immediately with the Connection Machine designers. Each simulated neuron can map to a single processor and the signals between neurons can ride along the network in the computer—an ideal application for the machine.

Soon after, the ever restless Feynman turned his attention to modeling quantum chromodynamics calculations. This is the subfield of quantum physics that relates quarks and gluons to protons and electrons. Performing calculations using this theory is a massive job involving huge matrices (tables of numbers). Feynman showed that the Connection Machine could do this even faster than a Caltech machine built explicitly for that purpose. The first Connection Machine became operational in 1985. Programmers and scientists have since used it for problems in database searching, geophysical modeling, protein folding, climate modeling, and even reading insurance forms.

For example, the oil industry uses the Connection Machine for acoustic reflection analysis. Surface explosions generate acoustic shocks from the surface of an area suspected of having oil. The underground reflections generate hundreds of millions of numbers. A Connection Machine takes these numbers and then produces an accurate picture of the geological formations. Companies drill only when the geology is promising. Massively parallel processing is clearly useful.

> Now that doesn't mean that you couldn't invent a problem that can't be done by a lot of slow processors working together. It turns out that such problems are surprisingly hard to invent. And I think that's one thing we discovered in applying the Connection Machine in a broader context. It's very hard to find a problem that has a lot of data where the potential for parallelism doesn't grow at least proportionately with the amount of data. The only problems that don't run well on a parallel machine are problems that just have a small fixed amount of data.
>
> For example, if you simulate the motion of the nine planets of the solar system, you can represent that with eighteen numbers. It's hard to see how you would use a whole lot of parallelism with only nine or even a hundred planets. I'm not saying it's impossible—I'm just saying it's hard to see (which means it is a good problem for someone to work on).

Inspired by the inherent parallelism of the brain, Hillis had changed computer technology forever. By the 1990s, there were about twenty massive parallelism projects in various stages of commercialization. The scientists involved in these efforts, including Burton Smith, the chief designer of the Tera machine, have watched the evolution of the Connection Machine with great interest.

Hillis has reached several conclusions from the commercial experience with Thinking Machines. First, it is better to have a relatively few fast processors than a million slow ones. Otherwise, the cost of connecting them together is too high. The fifth model of the

Connection Machine, the CM5, included up to a maximum of about 16,000 microprocessors, but each microprocessor can manage about 120 million operations per second, giving a total of a little more than 2 trillion operations per second.

Second, it is better to use high-volume microprocessors with an existing market than to try to make specialized microprocessors (as Burton Smith has done). Hillis has chosen industry-standard Sparc chips, but the Connection Machine design can evolve as other processors come along.

Third, the network remains of critical importance, because many instructions must access faraway memory chips. This insight led Burton Smith to avoid local memory and to make each processor work on hundreds of tasks at the same time. By contrast, the network of Hillis's most recent Connection Machine uses what is known as a *fat tree*, invented by Professor Charles Leiserson of M.I.T. The cleverness of Leiserson's invention is best explained by another analogy.

A normal communication tree is like a conventional corporate hierarchy with strict reporting lines. Communication between members of a group (ideally) works easily and fast. Communication from one division to another, however, can be more problematic. It requires going up and down the management chain through a common boss. If there is a lot of interdivision communication, each higher-level manager must convey more messages, so interdivision communication will be painfully slow.

A fat tree is analogous to having teams of coequal and autonomous bosses at each management level, with larger teams at higher levels of the hierarchy. This reduces the bottleneck to communication inherent in the classical design, because any member of a management team can route the message to the next higher or next lower level (see Figure 1). Recent Connection Machines use fat trees.

Silicon Evolution: Creating Artificial Life

Biology has inspired Hillis in his successful philosophy of computer design. His recent work has shown that computer science can return the favor. Of course, biologists already use computers for word processing, data storage, statistical analysis, and pattern recognition tasks.

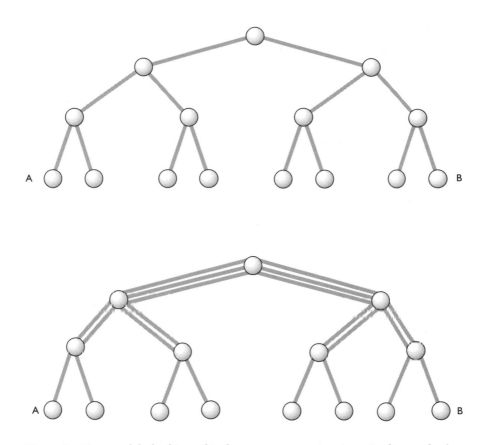

Figure 1   Two models for hierarchical corporate reporting trees. In the standard tree model, top, all directives from division A to division B must go up to one person at the top level and back down to the lowest level. In the fat tree model, bottom, there are several routes possible from division A to division B.

But their main business is running experiments on viruses, cells and higher level organisms and deriving models from these experiments.

Physicists—and to some extent chemists—take nearly the opposite approach. They often start with an abstract model and then propose an experiment to test it. What's more, physicists tend to believe elegant, abstract theories even in the absence of perfect experimental corroboration.

For example, in his book *Dreams of a Final Theory,* physics Nobel laureate Steven Weinberg observed that the first experiments to test Einstein's theory of general relativity gave results that only approximated the theory's predictions. The physics community reacted by assuming that the experimental apparatus was somehow faulty! (It turns out the physics community was right, but their faith in the theory came from aesthetic not observational considerations.) According to Hillis, biologists would never react this way.

> Biologists believe that simple mathematical theories are usually wrong, because biological systems are multicausal, poorly partitionable—basically, messy. Biological systems do have a beauty, but it is one of complexity and richness, rather than the simple reductionist elegance of physics.

If biologists are such a skeptical lot, why do they take a leap of faith and analogize from lower animals to higher ones? The best justification of evolutionary theory is that all known life is related by a common ancestry verified by the fossil record. Mathematical or computational models, by contrast, are constructed models of life-forms without any physical evidence. One never knows if a model's hidden assumptions might distort the analysis of the predicted phenomenon.

Just as Hillis in the 1980s believed the brain's organization might solve some difficult computer science problems, today he believes that simple mathematical models backed by powerful computers may lead to new biological insights. As an example of a mathematical model that enhanced our knowledge of biology, he cites the classical cybernetic notion of feedback, formalized in the 1930s at Bell Labs and M.I.T. Feedback is the idea of using errors in an output to adjust the input. After feeling hot water in the shower, feedback causes us to turn the faucet and produce a stream of cooler water.

> The brain isn't made up of operational amplifiers. But understanding the principles of feedback in amplifier circuits was very helpful in understanding how certain biological systems work. That became internalized in biology and is now something that is built into our understanding of how organisms work. We don't think of feedback as having come from cybernetics.

In his computer-based study of evolution, Hillis has two comparative advantages over traditional biologists. First, the incompleteness of the natural fossil record makes it hard to study the details of natural evolution. Second, experimental evolution, with even the most promiscuous fruit flies, is hard to achieve for more than a few hundred generations. By contrast, Hillis can run a simulated population for 100,000 generations on his Connection Machine and obtain a complete "fossil" record.

His basic technique is to construct a set of simulated organisms, each encapsulating a simple program. He then chooses a certain fitness criterion, such as the ability of the program to order (or "sort") numbers from lowest to highest quickly. He would reward the "fitter" programs by giving them a higher probability of reproducing in the next generation, albeit with some likelihood of mutation.

One of Hillis's early models showed that for some fitness criteria, sexual reproduction (mixing the programs of two "parents") outperformed asexual reproduction, but in other cases it did not. He then introduced parasites which brought in unsorted sequences of numbers and killed off programs that failed to order them quickly enough.

The parasites also evolved—their ability to devise tricky sequences of numbers determined their fitness. The surprising result was that the surviving sorting programs evolved more quickly toward the faster sorting solutions.

Right now, in the sorting competition, people still have a slight edge. Hillis's most evolved sorting networks are 2 percent more expensive than the best network found by a human researcher, Milton Green.

Oxford biologist William Hamilton had already suggested that parasites could help evolution. Hillis's model seemed to confirm that hypothesis. A cynical biologist might claim that Hillis hasn't added anything new, but simply designed an artificially confirming experiment. But Hillis's spare model has the advantage of exposing its assumptions in black and white—that is, in program code. Hillis's code shows that accelerated evolution will take place when three preconditions hold: parasites, the ability to mutate, and a fitness criterion tied in closely with a goal (fast sorting in this case). A question remains: Do these preconditions hold in the real world? Hillis confesses he doesn't have the answer.

I'm not sure that my notion that evolution solves a problem is right. Evolution doesn't solve a problem. Evolution invents a problem and solves it at the same time.

Hillis is only one proponent of "artificial life"—the study of the reproduction and evolution of computer-generated organisms. He and his like-minded colleagues, notably at the Santa Fe Institute, have found a cool reception among biologists. As John Maddox, the editor of *Nature* pointed out, "You can't expect people who have spent years just getting their organism to grow to have much respect for someone who does something on a computer in a few hours." Some biologists even go so far as to define artificial life into irrelevance. Hillis politely disagrees.

I asked [Nobel Prize–winning biologist] David Baltimore once why he thought this stuff wasn't life. He said to me, "Life has to evolve, has to reproduce, metabolize and be made of carbon." So by his definition, anything that runs on the computer is not going to be life.

Now to me the fact that it reproduces and evolves is much more interesting than it's made out of carbon molecules. But we never before had to distinguish between those two things—we never had an example of something that reproduced and evolved and wasn't made out of carbon molecules. I don't care whether we end up using the word *life* for this new thing or not. I think it's interesting whatever it is and it shares some properties with stuff that's made out of carbon and it's different in some ways. It doesn't prove anything in itself about the stuff that is made out of carbon, but it is suggestive.

Hillis's research on the Connection Machine combines his love of biology with computer science. Outside of work, he is an amateur geologist and likes to explore old mines. He's even staked out a claim in New Mexico, where he may move now that Thinking Machines has left the hardware business for financial reasons. He also designs toys—he did so professionally for Milton Bradley while in college and his creations fill his office.

Minsky and Feynman taught me how to think and they had very different ways of thinking. But one thing they had in common is a willingness to question everything. A lot of our shared reality is a bunch of shared unquestioned assumptions and both were good at pulling out those assumptions. Also, both Minsky and Feynman had a lot of fun at what they did. I think that's important. I think I do too.

A recent whimsical conception is a giant clock, the size of the Great Pyramid. The clock would tick once a year, bong once every hundred years, and the cuckoo would come out on the millenium.

It might make you wonder what the world will be like a few cuckoos from now.

$\mathcal{S}$CULPTORS OF MACHINE INTELLIGENCE

HOW TO MAKE MACHINES SMART

Part 4 of this book discusses research in a field peopled by some of the most innovative, if at times overly optimistic, researchers in computer science. Their dream is to make computers as smart, or perhaps smarter than people. Artificial intelligence (AI, for short) researchers want computers to act as if they can think the way people can: they should be able to recognize pictures, learn from experience, and reason in the irrational but adaptive way that we do so often. Among the products resulting from such a feat would be robots that clean houses and fix spacecraft and programs that practice medicine. Remarkably, the hardest tasks for people to master—medical expertise, in the above list—tend to be the easiest for computers.

And housecleaning presents a greater challenge than spacewalking. Deciding which objects to put in the trash requires the robot to distinguish objects from their backgrounds, recognize them, and use a form of commonsense judgment to decide what to do with them. These "simple" skills remain beyond the state of the art.

In fact, if we were to score modern-day computer capabilities by the standard of human-level intelligence, we would conclude that it's a toss-up: computers calculate better; have better memories for well-organized corporate minutiae; beat most people at chess; master arcane specialized knowledge; and are pretty good at graphics; but they fail to understand speech, recognize faces, or perform everyday reasoning at the level of a 5-year-old. Here we discuss the people who want the computer to succeed where it has so far failed.

After *John McCarthy* (whose profile is in Part 1 because of the importance of his work to programming languages) invented the programming language Lisp in 1958, he embarked on his fundamental quest: to teach computers common sense. The goal seemed simple enough, since it can, as he once wrote, "be carried out by any non-feeble-minded human." His efforts, however, have uncovered problems at every turn. To begin, common sense requires educated guesses. How much trust do you put in these guesses? They may turn out to be false. An external event may force you to revise them. Determining which events are relevant is an enormously difficult problem for a computer, even though people find it easy. Undaunted, McCarthy has laid out a logical framework for solving most of these problems, a framework that only now is being seriously tested.

Edward A. Feigenbaum has taken a more pragmatic approach. With his collaborators at Stanford University, he constructed the first programs to give expert advice based on a large body of knowledge in medicine (Mycin) and spectroscopy (Dendral). These programs showed how to construct expert systems and created a whole industry with applications as diverse as controlling aircraft flight plans to making loan decisions.

Douglas B. Lenat started his career in computer science with a program he originally called Automated Mathematician (but later shortened to AM). AM started with notions of mathematical sets, the ability to count the members of sets, and operations like set union, intersection, and difference. It went on to derive addition, multiplication, primes, and exponents. It even rediscovered Goldbach's conjecture: every even number greater than 3 is the sum of two primes. Although the program successfully reconstructed the basics of arithmetic and elementary number theory, Lenat saw it thrash around aimlessly as it struggled to discover more. Lenat decided that for a computer to discover new facts or ideas, it had to be primed with related facts and concepts. For the past decade, he has tried to encode the knowledge of a modern-day North America into a computer program called Cyc (from en*cyc*lopedia). The largest scale AI experiment ever to be attempted, Cyc may lead the way to extremely intelligent agents or may flop.

There is always a whiff of "it will never work" when one talks to most computer scientists about AI, but why shouldn't it work? More than 4 billion human computers are pretty good at it.

Edward A. Feigenbaum

THE POWER
OF KNOWLEDGE

There are three important things that go into building a knowledge-based system: knowledge, knowledge, knowledge. The competence of a system is primarily a function of what the system knows as opposed to how well it reasons.

—EDWARD A. FEIGENBAUM

xpert systems are part of our everyday lives. They run stock portfolios, help design cars, schedule airplane routes, and plan wars. Many researchers and practitioners have worked on expert systems, but one scientist has shaped the course of their development more than any other.

Edward Feigenbaum was born in Weehawken, New Jersey, in 1936. He never knew his father, an immigrant from Poland, who died just before Edward's first birthday. His mother remarried, and it was Feigenbaum's stepfather, an accountant and office manager in a small bakery, who would spark his interest in science.

> Once a month, the shows at the Hayden Planetarium in New York City would change. My stepfather would take me to each new exhibit. Afterwards, we would explore one little part of the Museum of Natural History. Eventually, after all these months, we covered the whole thing. That's my first recollection of scientific things.

An avid reader, Feigenbaum did well in school and excelled in science. He was enthralled by the massive, electromechanical Monroe calculator his stepfather used at the bakery.

> There were large table models that weighed about twenty pounds or more with lots of wheels inside and motors and a big keyboard. You'd punch in the numbers and you'd press big buttons that read "plus" and "multiply." Then the wheels would turn and go clank, clank, clank.
>
> I was fascinated by what the Monroe could do for me—it could do all these complicated calculations that I would struggle to do by hand. In order to impress my friends with what it could do, I carried it with me on the school bus going to high school. This thing was so heavy it was a major effort carrying it. And they were not impressed at all. So I carried the thing home.

A career in science appealed tremendously to Feigenbaum, but at his parents' urging he chose electrical engineering instead.

> There was enough attention paid during my young life to getting enough money for day-to-day living. Engineering seemed a more satisfactory alternative than science—more practical, more money.
>
> Electrical engineering was at the intersection point between science and mathematics. Everything in electrical engineering was relatively abstract. Whereas mechanical or civil engineering was the real stuff. I never was much of a "stuff" person. I'm more of a "thoughts" person.

Carnegie Years

When Feigenbaum entered the Carnegie Institute of Technology, now Carnegie Mellon University, in 1952, electrical engineers were still primarily concerned with power generators, radios, and the early televisions of the day. For the average undergraduate, computers didn't exist. Encouraged by a professor to look beyond the electrical engineering curriculum, Feigenbaum began taking courses at Carnegie's then new Graduate School of Industrial Administration. One of the professors there, James March, introduced Feigenbaum to the ideas of game theory developed by the Hungarian mathematician John von Neumann ten years earlier.

> That was fascinating—absolutely fascinating to me. That one could apply analytic and careful models to social phenomenon.
>
> For an undergraduate engineer, the science you're taught is pretty cut-and-dried. Newton's laws have been around for a long time. The laws of thermodynamics have been around for a long time. You learn to do the same kind of differential equations that a million other engineering students have done over time. All of a sudden, I was thrown in with a person who was explosively intellectual with wide-ranging ideas—all of them strange and wonderful.

At Carnegie, Feigenbaum also met Herbert Simon, a former administrator for the Marshall Plan and a political scientist by training. In his study of large bureaucracies, Simon had observed that individuals don't make rational decisions, rather they work from an implicit or explicit set of rules. He also observed that departments within organizations often focus on their own goals—for example, the shipping department wants to charge the maximum for delivery, even if customers complain. (Simon would later win a Nobel Prize in economics for showing the limits of rationality in organizations.)

In the early 1950s, Simon and Allen Newell, who had learned to program computers while working on an air defense project at the Rand Corporation, began discussing experiments using computers to simulate human thinking. They agreed to design their programs around the model that Simon had observed: a collection of components each having a subgoal and each driven by rules.

By 1955, when Feigenbaum was taking Simon's senior-level course on mathematical models in the social sciences, Newell and Simon had worked out a strategy.

> When we got back after Christmas vacation, Simon opened the class by saying he and Al Newell had invented a thinking machine. This was the Logic Theorist. We in the class said, "What do you mean, a thinking machine?"—that was a bizarre concept to us.
>
> Simon told us what he meant by a machine and handed out the manual to the IBM 701.

I took the manual home with me and read it straight through the night. When the next day broke, I was a born again—you can't say "computer scientist" because there was no such thing as a computer scientist at the time. Anyway, I realized what I wanted to do. So the next thing was, how to do that?

The Logic Theorist program attempted to discover proofs in the propositional logic of Russell and Whitehead's classic book *Principia Mathematica*. No one had ever written a program to make its own discoveries before.

Newell and Simon started by formalizing the idea of a heuristic. A *heuristic,* a term they borrowed from mathematician George Polya, is an explicitly stated problem-solving technique. For example, if you are trying to prove Y and know that X implies Y, then you can try to prove X instead. This particular heuristic is familiar to anyone who has taken high school geometry: When trying to prove that two sides of a geometric shape are equal, try to see whether the two triangles containing those sides are equal.

Then they specified a search strategy that consisted of removing a proposition from a global list of propositions and then applying all possible heuristics to it. Each heuristic might introduce a new proposition to prove, which would then be added to the list or replace a proposition on the list.

For example, if the proposition to prove is "Socrates is mortal" given the axiom that "every man is mortal," one can apply a heuristic to replace "Socrates is mortal" by "Socrates is a man" as the proposition to prove. This combination of heuristics and search would play a direct role in Feigenbaum's formulation of the expert system idea ten years later.

Feigenbaum stayed on at Carnegie to work on his Ph.D. with Simon at the School of Industrial Administration. He decided that he should also learn to program. In the summer of 1956 Feigenbaum went to work for IBM in New York City.

We young kids were down on the ground floor of a brownstone right around the corner from IBM's main building. They set us up with desks and everything and we were working away. Every once in a while we had a lecture by someone from IBM.

One day that summer, a guy comes down from the fourth floor of this building to tell us students about a marvelous new development that was happening. "You know how you students are sitting there writing codes called 'clear and add,' 'store,' and all that? You're not going to have to do that anymore. We're doing a thing that allows you to write formulas, and it's called Formula Translator or Fortran for short." By the fall it had come out. The person was John Backus.

By now, Feigenbaum decided that his research would be about computers. He wasn't exactly sure how, but certain dreams began to take shape.

> At that time, in the fifties, I didn't think of practical applications at all. I was intrigued by the vision of a highly intelligent, maybe super-intelligent artifact. That's why I sometimes characterize myself as a golem builder. Wasn't it the Rabbi of Prague who was building a golem? So maybe I'm a descendant of the Rabbi of Prague.[1]

Feigenbaum's doctoral thesis, however, was in psychology. In their work with Logic Theorist, Newell and Simon had modeled human problem solving. That is, they had tried to make a computer solve problems in order to see whether they could learn something about human problem solving. For his doctoral thesis, Feigenbaum would model a limited kind of memorization with more or less the same goals.

> Learning collections of rote materials in structured situations that were long studied in psychology laboratories: rote serial learning and rote paired associate learning. Trying to model the human information processing so carefully and so well that such a program put in an experimental setting would produce results that people would produce in that setting.

Feigenbaum called his program the Elementary Perceiver and Memorizer (EPAM); it is still in use today at Carnegie Mellon. The

---

[1] In the story, the rabbi creates a creature from a lump of clay who starts life as a servant, but eventually dominates his master.

program embodied a model for how people succeed in memorizing pairs of nonsense words, such as XUM-JUR and FAX-VUM, in a stimulus-response setting. In such a setting, the experimenter first presents a set of stimulus-response pairs to the subject, who attempts to remember them. That is called the *training portion*. Later the experimenter presents the first member of some pair and the subject attempts to remember the other member of the pair.

This learning process had presented psychologists with insights into the workings and capacity of short-term memory and the types of recall errors people are likely to make. For example, subjects are more likely to give the wrong response to a stimulus if that stimulus is similar to another one, and subjects are more likely to have trouble remembering a set of pairs after an intervening learning task.

Feigenbaum's objective was to present a computer model of memory organization that would explain subjects' incorrect as well as correct behavior. He called his memory model a "discrimination net." As training proceeded, the net would attempt to retain just enough information to distinguish among the stimuli so far received.

This is a strategy you might use when learning the jargon of a new field or a foreign language. You may hear the term *ROM* often, but remember it only as something having to do with a kind of memory in your personal computer. You then hear that *RAM* also has to do with some kind of memory in your PC and you're not sure whether RAM is the same as ROM or different. Only when you have heard about both of them do you make the effort to learn the distinction. Discrimination nets worked in the same way. In fact, they correctly predicted the errors that people made.

For the six years following his Ph.D., Feigenbaum, by then at Berkeley, continued to collaborate with Simon on extensions of EPAM and related projects. They published extensively in psychology journals. But Feigenbaum was still haunted by his dream of building a superintelligent machine.

At Berkeley, Feigenbaum was on the faculty of the School of Business Administration. There was no computer science department, and academic politics would keep things that way for some time. Berkeley's neighbor across the Bay, Stanford, had begun to blossom as a center for computer science thanks to a very supportive administration. George Forsythe was hired to get the computer science department going and he in turn hired John McCarthy to head the artificial intelligence laboratory. In 1965 McCarthy convinced Feigenbaum to join him at Stanford.

The First Expert System

I decided that I didn't want to be a psychologist and study human behavior using a computer. I wanted to build the computer artifacts. I moved from Berkeley to Stanford to be part of this new department.

In 1962, Feigenbaum had coedited with Julian Feldman a collection of papers on the state of AI called *Computers and Thought,* one of the most important books in the early history of the field. In the foreword, Feigenbaum advocated using the computer to explore processes of induction.

Artificial intelligence had been concerned with deductive kinds of things—proving theorems, making moves in chess. How about the process of empirical induction where we go from a wide variety of data to a generalization or a hypothesis about what that data means?

In 1878, the philosopher Charles Sanders Peirce explained the difference between deduction and induction using the simple image of a bag of beans. If you know a bag of beans has all white beans and you start picking beans from this bag, then you can *deduce* that the next bean picked from the bag will be white. On the other hand, if you know nothing about the bag, but you start removing beans from the bag and they are all white, then you may *induce* the rule that all the beans in the bag are white.

Induction, therefore, is a kind of informed guess, one that may have to be changed in light of new evidence. (Induction is similar to a nonmonotonic inference as characterized by McCarthy.)

So I began thinking. Newell and Simon had taught me that the best way to make progress on these difficult and vague problems is to choose particular tasks and work on them, as they did with their tasks in chess playing, proving theorems in propositional calculus.

> I spent quite a lot of time thinking about what would be the
> appropriate task to explore empirical induction. A couple of years
> went by without my nailing down that task.

Feigenbaum considered many possibilities—he even considered
testing whether a computer could induce the rules of baseball by
being exposed to events from a baseball game. But none of these
ideas seemed quite right.

Then in 1964, at meetings of the Center for Advanced Studies in
Behavioral Science at Stanford, Feigenbaum got to know Joshua
Lederberg, chairman of the genetics department and a Nobel Prize
winner. Lederberg suggested a problem that he was working on in
exobiology—the search for life on other planets. Lederberg was
working on a Mars probe that would land on the surface of the red
planet and explore for life or precursor molecules.

> A key instrument on this planet probe was going to be a mass spec-
> trometer. Lederberg's laboratory was doing measurements of mass
> spectrometry of life precursor molecules, particularly amino acids.
>
> When he mentioned that problem to me, I thought it was the
> quintessential empirical induction problem, just exactly what you
> would expect scientists to do. There's an array of data, and they get
> out an instrument and they think hard about it with all their ana-
> lytical tools and they come up with a hypothesis as to what is the
> best candidate to explain all that disparate data.

But Feigenbaum and his team realized after several months of
collaborating with Lederberg that they were missing essential knowledge about chemistry.

> The artificial intelligence processes we were using had already been
> developed. We didn't have to develop them further—they were
> search processes that were fairly well understood. What we needed
> was the knowledge of chemistry that would allow these processes
> to produce the right answer.

Lederberg was a world-class geneticist but not a chemist.
Feigenbaum and Lederberg convinced another Stanford colleague,
Carl Djerassi, to join their project. Djerassi, renowned for his work
in mass spectrometry and as one of the developers of the birth con-

trol pill, would provide the needed knowledge base for Dendral, as the project became known.

By 1965, the successful Dendral project could stake its claim as the world's first true expert system (see box below). It was able to determine the chemical structure of molecules sometimes even better than Djerassi's students, though it has not yet made it to Mars.

DENDRAL AND ITS DESCENDANTS

Dendral developed into a sequence of projects. The genesis was Lederberg's original 1964 algorithm, which, given a set of constituent atoms, generated all possible unringed structures. (An unringed structure is one containing no loops. They started with unringed structures for the sake of simplicity.)

The Feigenbaum-Lederberg collaboration of 1965, called Heuristic Dendral, provided a framework for applying mass spectrometry data and expert chemist rules to constrain this enormous set of structures.

A typical rule might relate the mass of a molecular structure to its spectrographic peaks in order to infer the existence of a certain structure, say a ketone group. Applying knowledge and heuristics would reduce the number of possible structures enormously, often by a factor of 200.

The program consisted of three parts. The *plan* stage identified molecular fragments that had to be in the final structures and fragments that could not be in the final structure. The *generate* stage created a list of possible structures consistent with the constraints of the plan. The *test* stage ranked the resulting list of possible structures by simulating the readings of a mass spectrometer on that structure.

META-DENDRAL

The next major advance came in 1970 with the program called Meta-Dendral. Its goal was to infer new rules about mass spectrometry for subsequent inclusion into Dendral. Given pairs relating known structures of known atoms to their mass spectra, Meta-Dendral would propose rules for inferring structure from the mass spectrometer data. Some of these rules proved to be brand new and have improved the performance of the original Dendral.[2]

[2]For more information about Dendral see Avron Bear and Edward A. Feigenbaum, *The Handbook of Artificial Intelligence*, vol. 2, Los Altos, Calif.: William Kaufmann, 1982, pp. 106ff.

Dendral in its several incarnations proposed not only a solution to a particular problem, but also a framework for expert systems in general. Each expert system has a set of data (in Dendral's case atomic constituents and spectrometry measurements), a set of hypotheses (possible structures), and a set of rules to choose among the hypotheses (relating spectra to structural constraints).

Whether this framework would apply to other domains remained to be tested. Feigenbaum believed it would—provided he could find the right experts.

Mycin, an expert system designed to help doctors diagnose infectious diseases and recommend treatment, emerged in the early 1970s from a collaboration with Stanley Cohen, Bruce Buchanan, and Edward Shortliffe. Cohen would later develop the fundamental technology for recombinant DNA. Shortliffe and Buchanan were the main implementors of Mycin.

A significant Mycin innovation was to include probability measures to weigh the likelihood of certain conclusions. A typical rule might relate the facts that an infection is bacterial, the culture site is sterile, and the portal of entry is from the intestines. It might conclude with a 70 percent confidence level that the organism is *Bacteriodes*. Thanks to the confidence factors, physicians could keep track of multiple hypotheses and rank them from most likely to least likely.

> What was nice about medicine was that it was a wonderful sandbox to play in—just full of heuristics. That's what medicine is. There is very little cut-and-dried knowledge. Very little black and white. The art of medicine, rules of good guessing, experiential knowledge, rules of good judgment. All mixed in with reasoning processes and of course, real data. You do measurements on patients, you do tests and then you try to figure out what the tests mean.
>
> Artificial intelligence methods are the methods of choice in domains of ill-formed problems. Many computational problems of the normal variety—from physics calculations to payroll—are well-formed problems for which algorithms exist. You don't have to do any search to solve the problem.

The Knowledge Principle

The world is full of ill-formed problems, but Feigenbaum decided that some were better than others. The key ingredient, he concluded, was an agreed-upon body of knowledge.

> Areas in which experts could articulate the basis of their choices, the basis of their judgments, were the proper areas for us to work in at an early stage.
>
> So, for example, it would be better to work in a medical area or an engineering area where we found people could be quite articulate about how they were thinking about problems versus, let's say, economics where there isn't really any well-defined body of knowledge that people are using to make an economic forecast.
>
> I think as the techniques have developed over the years and people have gotten more skilled and confident in applying them, they've been applied to problems that I wouldn't have dreamed of attacking in the seventies, including, for example, economic forecasting or foreign exchange trading. People are now willing to work with models that are what you might call *authored*. That is, if there isn't a widely agreed-upon body of experts, but a particular individual might have a well-articulated body of knowledge that others may not agree with, then representing that knowledge, and saying this is an expert system representing Joe Jones's knowledge, became a viable thing.

As of this writing, airlines use expert systems to determine how to deploy airplanes in various airports to meet demand during bad weather, mechanical difficulties, or airport congestion. In the case of American Airlines, each of its 550 airplanes remains on a single routing for an average of only three days at a time. Other expert systems help planes avoid collisions.

Busy European railroad stations such as Paris's Gare de Lyon use expert systems to route trains to various platforms with the goal of minimizing track crossings.

Computer hardware manufacturers use expert systems to decide which combinations of processors, memories, disks, networks, printers, and monitors make the most sense for a customer. For Hewlett

Packard, this means making a choice among more than 20,000 different alternatives.

Hospitals use expert systems to determine the odds that a patient in intensive care will survive, rating each patient on a scale of 1 to 100. It then suggests which patients should get how much of the hospital's resources. Not only doctors can play god.

Expert systems can also help generals.

> Here's one example: ARPA [the Advanced Research Project Agency, the Defense Department's research funding arm] asked some of its contractors to get very busy after Iraq invaded Kuwait to convert one of the ARPA-sponsored computer programs for manufacturing scheduling into a logistics scheduling program to help logistics officers plan the movement of men and matériel from the U.S. and Europe to Saudi Arabia. And they did it successfully.
>
> The director of the whole agency said that this one application had paid back the entire investment that ARPA had made in artificial intelligence since the beginning.

In Feigenbaum, one finds a unique combination of academic and business expertise. As a founder of IntelliGenetics and a director of Teknowledge and IntelliCorp, he has guided the development of expert systems in biogenetics, engineering, and general commercial use. The center he founded at Stanford University, the Stanford Heuristic Programming Project, has trained scores of graduate students who have carried their experience with them across the planet. They have been advocates for the widespread use of Feigenbaum's conception of an expert system.

> A new technology does not make it into the real world by itself. It makes it by virtue of the championing of that technology by some person or organization. Technologies don't sell themselves.

One thing his students have undoubtedly heard is Feigenbaum's belief in equipping expert systems with sufficient precise knowledge. He calls this the "knowledge principle."

> Suppose we are at Georgetown. Georgetown has a great medical school. It also has an excellent math department. So we are sitting

here and one of us gets sick—has a heart attack or something. We rush that person over to the hospital where there are people trained in medicine. We don't rush them over to the math department where there are excellent reasoners. Right? Because reasoning doesn't make any difference. You need to know about medicine, not about reasoning.

Until becoming chief scientist for the U.S. Air Force, Feigenbaum was codirector of the Knowledge Systems Laboratory at Stanford University, a position he used as a bully pulpit to advocate the broader use of expert systems and sometimes to admonish his colleagues.

I am thrilled with the fact that the technology has been widely deployed with such benefit. I am baffled by why a technology this powerful didn't become hugely successful.

Feigenbaum believes that there have been some mistakes in the marketing and presentation of many expert systems.

The field went around selling knowledge-free software shells [reasoning software]; they didn't sell the knowledge. The second thing was that we couldn't break the knowledge acquisition bottleneck. Knowledge acquisition is too hard. Working with the experts, getting knowledge from them, getting them to articulate their model of the domain is a difficult and time-consuming problem for which we had given the field no tools.

The knowledge principle says the performance depends on the knowledge base, not the logic. So if the knowledge is out of date, or if you do not routinely update the knowledge, the performance of the system decays over time.

When comparing expert systems to the ubiquitous database systems (which hold data in record format but don't try to reason about it), Feigenbaum again points to knowledge as the root cause for the difference in acceptance.

During the entire first generation of knowledge-based systems in industrial applications, nobody ever reused a knowledge base. There was no way you could take this piece of your knowledge about, say, the steel industry problem and use it in another problem. Nobody did it. They always constructed a new knowledge base for every new problem. That's not the case with database systems. You put together one database and different people use it for different purposes.

Doug Lenat's Cyc project (see the next chapter, on Lenat) is a logical embodiment of Feigenbaum's knowledge principle. Cyc attempts to encode the everyday knowledge of a late-twentieth-century North American adult (assuming there is such a thing) into a computer program. One of the main motivations is to create a reusable infrastructure of common knowledge from which one can build expert systems. Such knowledge would prevent a system from accepting data that is clearly absurd—say, a 3000-mile flight that takes 2 microseconds—and from drawing similarly absurd conclusions.

Feigenbaum travels extensively to act as a spokesperson for the knowledge principle. Between business trips, he indulges in his love of tropical islands, especially Bali and New Guinea. In California, he sings baritone in the Stanford University chorus. He has collaborated with several authors, including his wife, H. P. Nii, a computer scientist, on several books on expert systems and business.

Feigenbaum is confident that the expert system will gain its rightful place as an intelligent agent that can cooperate with people to solve some of the world's most challenging problems.

> Humans are superb problem solvers; superb learners; superb at coordinating functions of sensing and locomotion and problem solving into an integrated unit. However, computer programs can claim intellectual niches that evolution did not provide for us marvelous creatures.
>
> We have to think here's one intelligent agent, there's another intelligent agent and they have complementary capabilities. We have to design our systems so that both of these are working together to produce a better result.
>
> Twenty-five years elapsed between the invention of the reaper and its first use in farming. Twenty-five years elapsed between the invention of the telephone and the pay telephone call. It takes a long time.

Douglas B. Lenat

A TWENTY-YEAR BET

*How many people have in their lives a 2 to 10 percent chance
of dramatically affecting the way the world works?
When one of those chances come along, you should take it.*

—Douglas B. Lenat

ike artists, scientists often weigh the risk of ridicule against the risk of obscurity. The question, "Will people like this work?" takes on the concrete form of whether your papers will be accepted in good conferences, and whether people will treat you with respect, laugh at you, or just ignore you. One sure thing about Doug Lenat: the last possibility is out of the question.

Still a young researcher, Lenat has set out to build programs to solve the central problems of artificial intelligence: to make a machine learn and to instill it with common knowledge and common sense. He has done so in the true spirit of an explorer, forging ahead with only a lightly sketched map in hand. To some critics, he follows in the footsteps of classical AI to a cul de sac from which he will never emerge. Other critics mix skepticism with indignation that anyone should attempt to solve the central problems of AI before the theory has been completely worked out. Even we, as writers of this book, have felt the reflected heat of those attacks. "Why Lenat?" many have asked. You will soon find out.

Beyond the Hood

Born in Philadelphia in 1950, Lenat grew up there and in Wilmington, Delaware. His family owned a soda bottling business. In the school library during sixth grade, Lenat discovered Isaac Asimov's popular books about physics and biology. Science became an outlet for his curiosity about how the world worked. When Lenat was $12\frac{1}{2}$, his father died suddenly.

> I was superficially religious—Jewish—up until the time I was twelve-and-a-half. A few months before my Bar Mitzvah, my father died. It was one of those things in which you decide the world was inherently not fair and regardless of the objective truth about God, it's not worth doing these observances if bad things happen.

After his father's death, Lenat's family moved frequently and he found himself often starting over in new school districts.

> I was constantly being put in the lower classes—the so-called "hood" classes. Since they hadn't had me in the previous years, they weren't going to put me in the advanced classes. You constantly had to prove yourself instead of resting on context and circumstances. People in the good classes were expected to do well and didn't work very hard. The people in my classes were not expected to do well and you really had to work hard.

During this time, young Lenat turned to science as a form of solace. His talents showed. In 1967 he earned a finalist's spot in an International Science Fair on the basis of his description of a closed form definition of the nth prime number. Along with the other winners, Lenat received an all expenses-paid trip to Detroit. At the time, Lenat was disappointed about the location—the previous year's fair had been held in Tokyo—but the fair had other benefits. Contestants were to be judged by practicing scientists, researchers, and engineers.

Before that, the closest thing to a scientist I had met was my high school science teacher.

Lenat entered the University of Pennsylvania in 1968. The Vietnam War was at its height, and Lenat received a draft number low enough to make him think he might have to go to war. The uncertainty of those times convinced him to speed up his academic training while his student deferment lasted.

Lenat started college interested in physics and mathematics, but he had changed his mind by the end.

> I got far enough along in mathematics to realize I would not be one of the world's great mathematicians. . . . I got far enough along in physics to realize that in some sense it was all built on sand—that people were spending their lives doing things like finding mathematical solutions to things like Einstein's astrophysical equations—whether or not it had any physical significance or reality. . . . People would walk around with ever-growing chest pocket cards of elementary particles which really means resonances that were found but not understood. Things were just happening that divorced themselves from physical reality.

A course taught by John W. Carr III in 1971 introduced Lenat to artificial intelligence. Computers had created the technology for AI, but research, Lenat decided, was still at an early stage.

> [It was] like being back doing astronomy right after the invention of the telescope. The subject was inherently fascinating, but it had two other really interesting properties.
>
> One was that it was positively reinforcing—you would be building something like a mental amplifier that would make you smarter, hence would enable you to do even more and better things.
>
> The second interesting property was that it was clear researchers in the field didn't know what the hell they were doing.

Edward Feigenbaum and John McCarthy might object to the last statement, but it was the case that no one had built programs that seriously addressed many basic questions.

At the time Lenat was in school, the most successful applications of AI were rule-based systems to solve calculus problems (M.I.T.'s

Macsyma) or to simulate experts (Feigenbaum's Dendral), both completed in the mid-1960s. But this reflected a very limited form of intelligence. For example, it left out learning and everyday reasoning, though John McCarthy was working on the latter. As AI folklore of the day put it, "It's easier to simulate a geologist than a 5-year-old."

But Lenat wasn't discouraged, because the promise of mental amplification seemed to make it all worthwhile.

> Physics circa the 1800s and engineering even today have provided amplification to our *physical* selves. We can travel further and faster than we can walk. We can telecommunicate farther than we can shout.
>
> The AI goal is a kind of *mental* amplification so we can be smarter, be more creative, solve harder problems faster, forget less, be reminded of more.

Lenat graduated from the University of Pennsylvania in 1972 with B.A.s in math and physics and a master's in applied math. (By that time, U.S. participation in the war began to wind down, and Lenat's number was never called.) He declined an offer from Stanford and went to CalTech for a Ph.D., but after a month of reading the technical literature, Lenat decided that the AI action was really at Stanford and M.I.T. He called McCarthy to ask whether he could attend Stanford after all. McCarthy agreed, so Lenat packed up a U-Haul and drove from Pasadena to the Bay area.

As it turned out, McCarthy left soon after on sabbatical for M.I.T. Lenat started his work in computer science under a first-year assistant professor named Cordell Green. Green had done theoretical work in automatic programming (see box, "Automatic Programming") for his doctorate, using McCarthy's situational calculus. Upon arriving at Stanford, Green had converted to an experimental approach and like many an eager convert pressed that conversion onto Lenat.

> Before Stanford, I had seen myself as a formalist; Cordell (and my later mentors Feigenbaum and Buchanan) impressed upon me the value of being an empirical scientist even in an area like AI—looking at data, doing experiments, using the computer to do experiments to test falsifiable hypotheses.

AUTOMATIC PROGRAMMING

Automatic programming is the attempt to reduce programming effort. With automatic progamming, the user states a problem to the computer and the computer is then supposed to figure out what to do. For example, automatic programming would permit a user to present a set of tax laws to the computer and ask the computer to generate a program to minimize his or her taxes.

Although this goal has not been achieved, ideas from this field have inspired new programming languages such as Prolog and a family of languages known as *constraint logic programming languages*. Using these languages, programmers give a formal statement of a problem and its constraints. Given those constraints, the computer then tries to find a solution. The computer-generated program is usually much slower than a program that solves the problem using the best algorithms available, but researchers across the planet are working to shrink the speed difference as much as possible.

Failing in original goals but achieving other results is a typical fate of AI projects. AI researchers posed the questions and created the techniques that have shaped the apparently unrelated subdisciplines of database management, robotics, symbolic programming languages, and graphics. For example, robots were first constructed in AI laboratories, and applications such as motion planning and sensor-actuator coordination still use remnants of the original AI algorithms. Much of the past ten years of database management work has concerned efficiently incorporating expert system-style rules into database systems. What distinguishes such research from its origins in AI is the focus on efficiency and provable correctness.

Falsifiable hypotheses were introduced by philosopher Karl Popper to explain theories in natural sciences, but they play an important role in artificial intelligence. By his own account, Popper developed this notion while trying to resolve the question of how to prove a scientific theory. An experiment agreeing with the theory does not constitute proof, since a new experiment may undermine the theory.

After pondering the question for some time, Popper concluded that it is in fact impossible to prove a scientific theory. This insight led immediately to the question of how to distinguish a scientific theory from, say, a religious one.

According to Popper, an assertion is a good scientific theory if it can be tested by observation. An assertion that it rains when god X is angry is not testable, because one can never directly observe X to be angry. In other words, there is no conceivable outcome of an experiment that would disprove the god X assertion. By contrast, a theory that says that it is necessary to have clouds in the sky before it rains is testable: the first day you see rain without clouds renders such a theory false.

To Popper, then, a theory is scientific if it is *falsifiable,* that is, if an experiment could show the theory to be false.

Popper's views work beautifully for physical sciences whose theories are predictions about nature and therefore falsifiable. Newtonian mechanics was overthrown, for example, when the Michelson and Morley experiment showed that light doesn't play by the same rules as sound or balls thrown at moving trains. Experimental falsifiability doesn't work for mathematics, since that subject is really a set of tautologies: a statement ultimately reducible to A is A. For example, given the assertion that multiplying an odd number by three yields an odd number and the proof that goes with it, what falsifiability test could one rationally design?

Computer science lies in a strange middle ground. Theoretical computer science is essentially mathematics. Typically, one shows that an algorithm will take only so many steps of a certain kind for a given size of the problem. Such proofs reduce to mathematical tautologies, so cannot be tested.

Artificial intelligence programs, by contrast, interact directly with the external world, so they can be tested. A typical (though unspoken) falsifiable prediction is: "This medical expert system will tend to help more patients than it injures." But some researchers still create theories without seriously testing them.

For example, an AI scientist may propose a theory, then choose such a small test problem domain that his program can explicitly deal with its every conceivable outcome. Writing the program in this way makes the theory irrelevant.

The methodology of AI is still a sore point with many of its critics. Lenat sees the deficiency as a developmental problem related to computer science's short history.

> Computer science is really like physics in 1740. People are so happy for any results, that more than haphazard repeatability and duplication of other people's results are too much to ask. It's still at the stage where people are discovering rather than colonizing.

AM: Find the Right Heuristics

Lenat's first AI experiment was to write a Lisp program that formed concepts in mathematics (in Lenat's words, it "generated numbers and fooled around with them").

> Originally, [the name] stood for Automated Mathematician, but somewhere in the 1974–75 time period I got enough humility to let it just be called AM.

Lenat's AM program became his doctoral thesis and remains one of the most original AI programs ever written. As noted AI researcher Ernie Davis of New York University puts it, "It is completely unlike anything else I know of: a program with no inputs and no interactions that simply sits and considers mathematical concepts and proposes hypotheses forever."

AM's main scientific contribution was to embody the notion of learning heuristics. As you may recall from the Feigenbaum chapter, a heuristic in computer science is a problem-solving method which doesn't promise the best solution—it is a kind of educated guess—but which is often quite good. AI uses heuristics often, because AI programs must handle situations in which the program has incomplete information. In expert systems, heuristics embody a professional's knowledge applied to a particular situation. AM used heuristics to discover new knowledge. Some of these heuristics even apply outside mathematics. Here are some typical ones:

1. If an operation applies to two objects of the same type, then have it apply to the same object twice.

AM applied this to discover the concept of squaring from multiplication. Multiplication applies to two numbers—squaring is multiplication of a number times itself. In a similar way, doubling comes from addition. The value of this heuristic in mathematics is indicated by the fact that mathematics has special terms for many concepts so derived.

Even outside mathematics, concepts that embody this heuristic have special names. For example, from killing one gets self-killing or suicide; from analysis, self-analysis; from war, civil war.

2. Look for extreme cases.

AM applied the operation "divisors-of" to try to find numbers with zero divisors (there were none), numbers with one divisor (the number 1), and numbers with two divisors (prime numbers).

Extreme values often have special names outside mathematical domains as well. Consider the two extremes anarchy and totalitarianism in political thought; absolute zero in physics; and gibberish in speech.

3. If two concepts arrived at independently turn out to be the same, then explore this further.

AM used this to perform further explorations with the number 1 as a number with a single divisor and as the multiplication identity. Many people use this heuristic to decide which movie to see or which car to buy—one good review is good, but a good review and a good report from a friend are even better.

> The idea there was to put in a bunch of definitions and concepts, to put in a bunch of heuristics for judging the interest value and then just stand back and see what it discovers.

AM was based upon 115 set theory concepts and 243 heuristic rules. Working from this foundation, AM was able to "discover" 300 mathematical concepts. Each discovery involved from 20 to 50 heuristics, including the three described above as well as concepts of analogy. Besides discovering most of sixth-grade arithmetic, AM made conjectures such as "every even number greater than 3 is the sum of two primes." Making this conjecture—which was already known, as Goldbach's conjecture—was an impressive achievement for a computer program. On the other hand, the program didn't find what Lenat had thought it might.

> We did make up some scenarios ahead of time just to see if this was plausible. But it never did the scenarios that we thought of. We thought it would discover different kinds of infinity and Cantor sets. It never discovered that stuff.

AM aroused the curiosity of many of Stanford's leading mathematicians and computer scientists, including George Polya. Polya had written books maintaining that the best way to teach mathematics is to teach children to discover it.

Polya learned about AM through one of its results. AM had started exploring the smallest numbers with many divisors. For example, the first number with six divisors is 12—1, 2, 3, 4, 6, and 12 itself. AM tried to find the first number with seven divisors, eight divisors, and so on. Lenat wondered whether any mathematician had ever thought of such a thing. Polya seemed to be the only one who knew.

> He said, "This looks very much like something the student of a friend of mine once did." Polya was about 92 at the time. It turned out the friend was [Godfrey] Hardy and the student was [Srinivasa] Ramanujan. Early in the century, Ramanujan had come up with something about highly composite numbers very similar to one of the patterns that AM had discovered.

There were other Stanford luminaries who monitored AM's progress, including Donald Knuth.

> Knuth used to pore through the outputs of AM as if he were going through Babylonian or Aztec scrolls or hieroglyphics—looking for things that I and the system didn't realize were valuable.
>
> Every now and then he would find something that AM had come up with and didn't realize the significance of (and neither did I) and he would circle it—that was wonderful.

AM had been a success, but only within a limited domain in mathematics. As the program moved further away from elementary set theory, it began to "thrash" (Lenat's word for wasting its time by going down useless search paths), and the heuristics lost their power.

After completing his Ph.D. at Stanford in 1976, Lenat went on to Carnegie Mellon as an assistant professor determined to put the systematic study of heuristics at the center of his work. At Carnegie Mellon and then at Stanford, to which he returned after two years, he did his second major experimental program in artificial intelligence, called Eurisko.

Eurisko's goals were to discover new heuristics, as opposed to just using the ones it was given, and to create journal-caliber mathematics.

Instead, Eurisko's principal successes were to complete some innovative circuit designs and to construct a winning fleet for a naval war game contest called the Traveller TCS war games. The program used its heuristics to find odd loopholes that the game designers had missed. For example, a fleet would sink its own damaged ships to increase its mobility. After he won in 1981 and 1982, the contest organizers barred Lenat and his program from the contest.

But the program was still difficult to extend to other domains. From his experiences with AM and Eurisko, Lenat concluded in 1983 that even the cleverest heuristics were not enough.

> Learning occurs at the fringes of what you already know; so that you learn some new things similar to what you know already. If what you are trying to learn is not too far away from what you already know, you can learn. The bigger that fringe is—the more you know—the more likely it is, the more possible it is, to discover new things.

Cyc: Priming a Computer with Common Sense

If learning depends on knowledge, Lenat reasoned, then if a computer program is to discover something really interesting, it must know as much as people know.

> The idea was to stop for ten years; put in the knowledge that was required to prime the pump adequately, and then in the mid-1990s go back to machine learning.

Lenat wanted nothing less than to instill into a computer program the knowledge of a well-informed North American adult. From that base, he thought, the program could do some interesting learning. The name of the projected program was Cyc.

> Alan Kay, Marvin Minsky, and I got together and did some back-of-the-envelope calculations—we actually killed about five minutes to find an envelope so we would later say we did back-of-the-envelope calculations— on how much knowledge would be required and how much time it would take. That was a million frames over ten years.

A *frame,* an invention of Minsky's, is a way to represent real-world objects inside a computer. For example, the frame for a car may state that a car is a physical object about 4 feet high, between 6 and 40 feet long, having wheels, a motor, seats, and mechanisms for steering, accelerating, and braking.

Entering a million mostly handcrafted frames is an enormous job, and the three scientists quickly concluded that a project of this size would require correspondingly enormous financial resources. This would come from an unusual source.

Back in 1975, while still a graduate student, Lenat had attended the International Joint Conference of Artificial Intelligence (IJCAI-75) in Tbilisi, then part of the Soviet Union. It was a memorable event for two reasons. It was Lenat's first public talk, and he remembers his mortification when the high-powered blower on the Soviet slide projector scattered his viewgraphs to the four corners of the stage. But another event would prove more significant to Lenat's career: he met Woody Bledsoe. During the conference the two scientists had several animated conversations about Lenat's work. Eight years later, in 1983, Bledsoe became head of AI for the Microelectronics and Computer Technology Corporation, MCC, a new company being formed in Austin, Texas. He arranged for Lenat to talk to the head of the outfit, Admiral Bobby Ray Inman.

Before entering the computer business, Inman had enjoyed a long and prominent career in the government intelligence community. He had served in various naval intelligence functions and then as a director of the National Security Agency and finally as deputy director of the Central Intelligence Agency. (He nearly became Secretary of

Defense in 1994, but then withdrew, complaining of attacks by the press.)

> Inman was the most charismatic person I'd ever met. He was very good with people and making you believe that he was letting you in on the real story, the straight story. People would come to him raging at one another and they would walk out of his office, each one of them believing that they had won the argument.
>
> He really had this magical ability and everyone at MCC adored him. We would have meetings every few months about MCC and it would almost inevitably drift into world affairs because you would be so interested in what he thought about the world situation. It was Camelot.

In computer science, as in other scientific fields, funding is destiny. Many of the great advances in computer science have been spawned by the R&D departments of major corporations. Backus developed Fortran at IBM Research while UNIX came from Bell Labs. Academic researchers receive most of their funding from government scientific, commercial, or military agencies.

> Inman said, "This is the only place [MCC] and the only time in history that you will be able to do a project like this—a project that has a 2 percent chance of succeeding. You're never going to be able to do this with government funding, you're never going to be able to do this in academia. A single company, even a multibillion-dollar company is not going to spend tens of millions of dollars."

Not only would MCC and its shareholder corporations be committing substantial resources, but Lenat figured he would be spending a good percentage of his working life on a project that might not succeed.

> Ed Feigenbaum's influence on me: you figure that as a researcher you will have on the order of three decade-sized bets to make in your life. So, you might as well make each one count.

If Lenat were to succeed with endowing a computer with the knowledge of an average adult, the benefits from the program's applications would be enormous.

This would basically enable natural language front-ends and machine learning front-ends to exist on programs. This would enable knowledge sharing among application software, like different expert systems that could share rules with each other. It's clear that this would really revolutionize the way computing works, if it worked. But it had an incredibly small chance of succeeding.

It had all these pitfalls—how do you represent time, space, causality, substances, devices, food, intentions, so on. Things which AI researchers had gotten trapped in over the last thirty years, and philosophers had gotten trapped in over the last thirty centuries—giving their lives to the little nuances.

After experimenting with different ways of representing knowledge about the world in the project's first few years, Lenat decided that there couldn't be just one representation for each knowledge chunk. For one thing, knowledge depends critically on context. Fortunately, a young and talented Ph.D. student, R. V. Guha (a student of McCarthy's and Feigenbaum's), had worked out a model of "microtheories."

For each microtheory of a human activity such as shopping or driving a car, you would need to be able to make different assumptions and decisions.

Essentially, you have to give up the notion of global consistency and live with local consistency. Each theory, each context should be consistent, but there are inconsistencies across theories.

For example, suppose I say, "Who is Dracula?" and you say, "A vampire." Then I ask, "Are there vampires?" and you say "No." The contradiction doesn't bother you at all. The first question is being answered in a context—the Bram Stoker fictional universe. The second question is being answered in the everyday, scientific-world, skeptical context. You have no trouble separating these even though they seem directly contradictory.

Cyc uses microtheories to make assertions that are true in specialized contexts. For example, as a part of its microtheory of the "naive physics" of the everyday world, Cyc asserts that unsupported objects fall. A more sophisticated microtheory might contradict this rule in certain cases, e.g., in the presence of a levitating magnetic field or if the object is a helium balloon. Lenat believes that microtheories are essential to understanding metaphor and that metaphors are vital to our day-to-day existence.

> You couldn't really understand language if you didn't understand metaphors, even if you restricted yourself to a narrow corpus like *New York Times* articles about one American company buying another: there are predator-prey relations, first dates, seesaws, and warfare metaphors. I always liked in high school, you'd see headlines like "Our Mother of Mercy Slaughters St. Catherine." You need to have that richness and breadth of knowledge that percolate into almost all sentences.
>
> The critical thing is that most systems that have tried to be large have failed because they have tried to have global consistency: to decontextualize all the different assertions, to write down all the different assumptions. You can never do that. Each context has assumptions written down, but there are always some you've missed.

Thirty people under Lenat's direction at MCC were charged with entering knowledge—knowledge about thousands of everyday activities like shopping, football, visiting doctors, and so on. Lenat's team had to be aware of the context of each of their knowledge entries. But they also had to find a way to represent relationships between facts and beliefs. This involves subtle concepts of logic, such as *modal operators*.

> Suppose I say that Bill Clinton doesn't know that my age is 42; and my age is 42. Normally you could substitute equals for equals. But then the statement would be Bill Clinton doesn't know that 42 is 42. Whenever you have modals like "knows," "believes," "intends," "desires," you cannot replace equals by equals. They have to be treated specially.

By adding modal operators, Lenat and Guha greatly expanded the complexity of the machine's reasoning. Conventional problem-solving programs had been limited to having one or a few inference engines (rules for drawing conclusions). Cyc, by contrast, evolved into a system with 30 different inference engines.

The use of a multitude of inference engines enables the program to manipulate complicated, intangible concepts. For example, if a corporation owns a building, it also owns pieces of that building. The program would include the rule: "ownership" "transfer through" "physical parts."

Using many different methods of inference can lead to inconsistency, but Lenat believes there is no other way.

We avoided the bottomless pits that we might have fallen into by basically taking an engineering point of view rather than a scientific point of view. Instead of looking for one elegant solution, for example, to represent time and handle all the cases, look instead for a set of solutions, even if all those together just cover the common cases.

I have a kitchen sink ideology that says, if it works, we'll use it. We're willing to do anything to achieve this goal of mental amplification. . . .

It's like you have 30 carpenters arguing about which tool to use and they each have a tool—one has a hammer, one a screwdriver, etc.

The answer is they're all wrong and they're all right. If you bring them all together, you can get something that will build a house. That's pretty much what we've done here.

Cyc and Its Critics

Any project that lasts ten years will have its goals and methods modified. Although Lenat views this as natural, critics of his approach see it as a result of a fundamental weakness in his "scruffy" methodology, in contrast to "neat" theoretical formalisms.

Indeed, Lenat has a celebrity's ability to attract controversy. Debates about his work appear in the pages of academic publications and professional conferences. Two basic criticisms stand out.

1. Cyc ignores many theoretical problems concerning the representation of partial knowledge and beliefs and doesn't use some of the new, possibly promising, formalisms that have arisen since Lenat started his effort.

2. If Cyc fails, it will be unclear what lessons to draw from it. The designers have made so many choices, mostly without explicit rationale, that it will not be clear which led to failure. Moreover, if Cyc fails, it will be bad for AI. People will say that AI can handle only toy problems or narrow ranges of expertise.

Lenat responds to the first criticism by saying that he knows of these developments, but refuses to "cower in a dark corner until the theoreticians finish their debates." He goes on to say that new theories about AI are less important than sweat.

As for the second, Lenat points out that similar things were said about other big projects such as the Manhattan Project or the Space Project. He also says that he has no intention of failing.

Sometimes the critics are sympathetic. One reviewer notes that the scale of the effort demands a willingness to take risks and even to make wrong decisions. A beneficial side effect is that the Cyc effort has had to develop technologies that will be useful in any comparably sized effort.

If Lenat is concerned about his academic critics, he does not show it. His current goal is to make the project commercially viable. Several major corporations have helped to finance Cyc with half a million dollar per year contributions. These include Bellcore, Apple, Kodak, DEC, AT&T, Microsoft, and Interval Research. According to Lenat, the companies hope to use Cyc for very down-to-earth applications such as car-buying guides and smart spreadsheets.

Imagine explaining to Cyc the approximate meaning of rows and columns on your spreadsheet. Age, object bought on a certain day, annual salary, and so on. All of the things that you and I know about, Cyc has that level of understanding also.

Once the spreadsheet schema has been set up, then all of Cyc's millions of rules became applicable.

For instance, suppose a cell in a spreadsheet representing annual income gives the annual income of a third-grade teacher in Iowa as $25; it couldn't possibly be $25 a year—it has to be $25,000. You could do the same kind of data cleaning for relational databases. Say a person's record indicates that she is the mother of two people whose birth dates of record indicate they were born one week apart.

Lenat envisions other commercial applications of Cyc, such as photo archive searches.

> Suppose someone is doing an article for *Playgirl* and wants pictures of "shirtless young men in good physical condition." One of the candidate images is captioned, "Pablo Morales winning the butter-fly in the 1988 Olympics." How do you go from that query to that caption? You have to know that you are in some swimming event and intentionally in the water and you're probably wearing a swimming suit and if you're a male wearing a swimming suit in modern Western attire, you're shirtless. And Olympic athletes are in good physical condition. It's not a very long inference chain.
>
> But there's no way that using a thesaurus you could go from shirtless to the 100-meter butterfly. You either have the knowledge or you don't. If you have the knowledge, it's a trivial problem. If you don't have the knowledge, it's an impossible problem.

Lenat believes that Cyc's inferences could also be useful in medicine twenty years from now. Alzheimer's patients would be helped to remember things they used to know. Many of Lenat's speculations about Cyc and the future of AI sound like the science fiction that nurtures his adult imagination.

> You have to figure that some large fraction of science fiction is un-knowing fantasy—it's not meant to be but it is. Another fraction of it is unrealistic for various reasons. And one of the useful things to do is to think how it could be so completely wrong. Often this leads me to think about how we [in AI] should be doing something differently.
>
> For example, there are a lot of stories in which natural lan-guage understanding will come long before machine intelligence. In those stories, it's always a lot easier to understand language com-mands than to have common sense and wisdom and so on. If you think about how that could be wrong, you could say that maybe wisdom and common sense are necessary to understand natural language. And you can come up with a lot of examples in which that is true; you won't really be able to disambiguate the words in

INFERENCE CHAIN FOR THE PLAYGIRL EXAMPLE

Lenat sent us an annotated transcript of Cyc's actions in matching the query "Show me an image of shirtless young men in good physical condition" to the image having the caption "Pablo Morales winning the men's 1992 Olympics 100-meter butterfly."

Here are the essential steps of the transcript, without the Lisp code.

1. An existing axiom says that if X won event Y, then X participated in event Y. From this, Cyc concludes that Pablo Morales participated in this event.

2. Another axiom says that all Olympic men's events are instances of men's sports competitions.

3. Another axiom says that participants in sports competitions are young and have an athletic physical build; this is just a default, as are almost all the axioms in Cyc. Another axiom which is even stronger than that one—but more specialized—says that this is known to be especially true in the case of Olympic events. From this Cyc concludes that Pablo Morales is a young person and also that his physical build is athletic.

4. Another axiom says that men who are swimming wear swim trunks. From this Cyc concludes that during the swimming event, Pablo Morales is wearing swimming trunks. Notice that this still doesn't mean he is shirtless; that comes next, from the following axiom.

5. Another axiom says that if Cyc can't infer (guess) that a person is wearing X, assume he isn't wearing X. (Cyc is using McCarthy's notion of circumscription here.) Cyc then concludes that in the context of the video clip, Pablo is a young person, a male person, is physically in good condition, and is not wearing a shirt.

a sentence or translate this into that unless you really knew an awful lot about the world.

There are a lot of science fiction role-playing systems like Dungeons and Dragons. If you look into manuals for those systems, you can get a lot of insight into common sense. They have to make rules for everyday things.

Lenat has also drawn inspiration from several mentors over the years. One was Richard Feynman, the Nobel prize–winning physicist whose unstudied playfulness has been a model to a generation of scientists.

> I still think that the Feynman lectures are one of the pinnacles of nonfiction writing. He and I got along amazingly well from day one. He was very irreverent; he had a great sense of humor; he went out of his way to tweak people's noses if he thought they were doing things because of power or authority.
>
> I met him at Thinking Machines in 1983. We kept in touch but I only saw him a few times. We would just have wonderful bizarre bets—if you were growing an oak tree, could you turn it into an oak shrub simply by adjusting its water intake? He really liked what I was doing in research and I really liked what he was doing.

Lenat has also been strongly influenced by MIT's Marvin Minsky and vice versa. (Cyc is treated as a working reality in the futuristic novel Minsky wrote with sci-fi writer Harry Harrison called *The Turing Option*.)

> Even though Marvin and I don't do the same thing, we're fascinated by what the other is doing; each of us is a little worried that the other is more right than we think they are. Which means that part of what we are doing is wrong.

Like Minsky, Lenat believes that as artificial intelligence evolves, we will radically revise our ideas about what being human is.

> Up until now, people have lived linear finite lives. You exist in one place for a time, live for awhile, and then you die. That's going to change. Suppose that you have lots of intelligent agents standing in for you in cyberspace; all sorts of activities involving things you'd like to do. Your actual consciousness resides in one flesh body. Suppose that communication bandwidth is high enough that at any given moment you can shift your attention and inhabit these other entities. After a person died, some of his or her agents would still continue to be around. Some of Fred is around after Fred is dead.

If this future vision of a cyberclonic paradise were to come true, one Lenat entity would be able to scuba dive or parasail (virtually) near the Great Barrier Reef while another Lenat could scan an Isaac

Asimov book to ferret out the essential data of Gaia and several other Lenats would be playing games.

In the much nearer future, Lenat believes that AI projects like Cyc can become "knowledge utilities." The average person will have access to Cyc-like programs on personal computers at home, work, and school. Smart home marketing programs will help consumers weed through product recommendations or choose a movie, all based on an evolving computer profile. ("Information about information" is an industry known for profits. *TV Guide,* for example, is more profitable than most networks.) Knowing these preferences, companies would have another way to monitor erratic credit card purchases that might indicate that an owner's card had been stolen. If this sounds frightening, Lenat isn't scared.

> It's about time now to build on top of Cyc with machine learning. It will be as continuously creative as people are. My decade-long journey in the wilderness is almost over. I expect to spend a decent fraction of my time on machine learning. That's why I took this hiatus of ten years. If that succeeds, after another decade or so, I'll go back to general intelligence amplification, and who knows, perhaps then apply myself once again to physics or math armed with this "mental amplifier." Finally, at age 65, I'll be back doing the stuff I put on hold when I was 21.

Will Cyc work or won't it? The most likely eventuality is that it will work, but only in part. Cyc may render databases cleaner by eliminating obvious absurdities. It may also help consumers filter out irrelevant advertisements; for example, if Alice has just broken her leg, she probably is not now interested in buying running shoes. But Cyc may never be able to carry on a convincing conversation.

In the longer term, the mechanisms of Cyc—the microtheories, the interacting inference engines, the mechanisms for discovering new heuristics, and the recognition that AI requires lots of knowledge—will change AI research forever. If Cyc succeeds even partly, Lenat will have shown that AI demands sweat as much as brains. Thousands of tiny facts about the world are more important than a new theory of knowledge representation.

Why Lenat? Because the study of intelligence requires explorers of the boldest kind.

Epilogue

SECRETS OF SUCCESS?

What qualities of personality and intellect make great computer scientists? Judging from the fifteen individuals interviewed here, one might conclude that there are more differences than similarities.

Did their talents show at school, for example? Sometimes. Levin won first place in the Physics Olympiad for the entire city of Kiev as a teenager. But Backus recalls flunking out of high school every year. His parents sent him to summer school where he spent his time sailing.

Alan Kay continues to be inspired by his bad experience at school. He has spent much of this adult life using the computer to help kids learn. Burton Smith, the designer of the Tera machine, remembers constructing contraptions to fire spitballs during junior high school science classes. When 12-year-old Michael Rabin got kicked out of class for mischief, he found a career in mathematics: "There were two ninth-grade students sitting in the corridor solving Euclid-style geometry problems. There was a problem they couldn't solve. They challenged me and I solved it."

Three of our scientists, Kay, Knuth, and Lamport, came close to becoming professional musicians. Most of the rest enjoy music as a hobby. Burton Smith, whose designs allow hundreds of computers to work together, sings polyphonic Renaissance music in a choral group. But Danny Hillis, also a computer architect, likes poetry and rhythm but not music. Rabin claims that music is too hard—unlike mathematics or physics.

People make much of the fertile creativity of young mathematicians and the implied stagnation of middle age. This does not seem to hold for computer scientists. Rabin invented randomized algorithms in his forties; McCarthy invented nonmonotonic logics in his fifties. Backus worked on functional languages and Dijkstra developed new methods for mathematical proofs in their sixties.

On the other hand, early passions are often echoed in later accomplishments. As a child, Danny Hillis cultured a beating frog heart in a test tube. As a teenager, he became fascinated by neuroanatomy. Those interests led naturally to the design of the massively parallel Connection Machine and his later work in artificial life, getting the computer to simulate biological evolution.

Some computer scientists nurture their spirituality. Frederick Brooks credits a renewed Christian faith with his decision to leave a heroic IBM career (as project manager of the fantastically successful System 360) to found the computer science department at the University of North Carolina. Since then he has played a prominent role in piecing together the technology for virtual reality in the service of chemists, architects, and doctors. He feels a moral opposition to using virtual reality primarily for entertainment. In his retirement, John Backus reads the mystical writings of Eva Pierrakos. He says, "By looking into yourself you really get an appreciation of the mystery of the universe. You don't by trying to find the laws of physics."

The majority of our scientists profess less conscious spirituality. But they do remember moments of insight with a certain wonder. Dijkstra discovered his shortest-path algorithm while sitting with his wife at an outdoor cafe on a Saturday morning. Suddenly, he fell silent, realizing he had the solution. (He says of his wife's reaction to his sudden silence, "My wife knew such periods.") Burton Smith invented a hardware design for parallel machines while sipping a Scotch on an airplane.

While scientists in the Middle Ages found patrons in kings and princes, computer scientists depend on corporations, governments, and universities. IBM funded Backus and Rabin; Xerox and Apple supported Kay; an industrial consortium, MCC, has paid for Lenat's work. ARPA, the research arm of the defense department, or the U.S. National Science Foundation has funded everyone.

The type of funding matters almost as much as the amount. McCarthy and Minsky started an eleven-person research lab in artificial intelligence thanks to a casual conversation in the hallway with lab director Jerry Wiesner. McCarthy notes, "I think that such flexibility is one of the reasons the U.S. started in artificial intelligence ahead of other countries." In the course of the next year and a half, the group developed Lisp, the reigning programming language in artificial intelligence and one from which every modern language has borrowed ideas. Open-ended "we-fund-people-not-projects" grants allowed controversial ideas like Lisp, the Connection Machine, Smalltalk, and randomized algorithms the time to bear fruit. Alan Kay recalls how vague Xerox's goals for their Palo Alto Research Center were: "The only sentence I remember is if someone gets rid of paper in the office environment, it should be us."

Few computer scientists work in isolation. First, they go to great schools, either as students or professors. M.I.T., Stanford, Carnegie Mellon, Harvard, and Princeton pop up in the biographies of almost all our North American computer scientists.

Second, computer scientists often work together in small groups: Robert Tarjan and John Hopcroft; Michael Rabin and Dana Scott; Donald Knuth and Edsger Dijkstra; Alan Kay, Dan Ingalls, and Adele Goldberg. The collaboration of Rabin and Scott would strike many computer scientists as ideal: "One of us would formulate a question and then we would go to our respective corners and the other one would come up with a solution. Maybe overnight." A few, however, do work alone. Dijkstra credits "the happy fact of my isolation" with why he solved so many fundamental problems while others studied more fashionable, but ultimately less significant, questions. Lamport thinks he might never have obtained his results in distributed computing as an academic. "I didn't have to worry about tenure. I had the freedom to work on what I thought was interesting, not what other people thought was interesting." But isolation has its drawbacks. Sequestered behind the Iron Curtain, Leonid Levin saw his fame eclipsed because no one in the West knew about his results, and he didn't know about the independent discoveries of Cook and Karp.

Our computer scientists often made a stopover in physics. For example, Lamport's work on distributed systems draws direct inspiration from his ten-year obsession with special relativity. Four of the scientists knew Nobel prize-winning physicist Richard Feynman well. Lenat recalls, "We would just have wonderful bizarre bets—if you were growing an oak tree could you turn it into an oak shrub simply by adjusting its water intake." Feynman even went to work for Danny Hillis at Thinking Machines after labeling as "dopey" Hillis's idea of designing a brainlike machine by connecting a million processors. Playful, creative, and willing to try nearly any idea at any time, Feynman personified how computer scientists like to see themselves.

The widely shared interest in biology surprised us. Temperamentally, the two fields seem so different. Biologists experiment with wet organic material and distrust elaborate theories; computer scientists type at antiseptic silicon and metal–plastic sculptures and depend on a huge body of mathematical and linguistic theory to do their work. Yet both Hillis and Kay studied biology at the university level. Even Levin and Rabin, theoretical computer scientists who grew up breathing physics and mathematics, look to biology for the next great advances in computer intelligence. As Levin puts it, "Genetic history has been very short since we were jumping from tree to tree. Nevertheless the same algorithms we developed for catching bananas are good for harnessing nuclear power."

Any theory about why these scientists are so brilliant runs into the luck factor. None of these scientists won the lottery, but they share the good fortune of working at the right place at the right time. At his wits' end about which career to follow after finishing college at 25, Backus took a tour of IBM corporate headquarters and happened into a job. The frustrations of machine level programming led him and a few colleagues to think of a better way of doing things. McCarthy's invention of Lisp followed from his difficulty in using Fortran to solve AI problems. Feigenbaum studied with Newell and Simon. Kay's first job as a graduate student was to make Simula work.

Our scientists' own explanations of success vary fifteen ways. They agree on a few points, however. First, the problem you choose can determine the importance of your solution. Fred Brooks looks for problems that people in other disciplines come up with— chemists, doctors, and architects in his case. He calls them "driving problems." As he puts it, "It keeps you honest when someone can tell you, 'What you're doing is just as pretty as you please but it isn't any help to me at all.'" Finding the right problem can take time. Edward Feigenbaum, the creator of the first true expert system, spent three years considering several problems in induction, including the rules of baseball, until he decided on biochemistry as the best candidate for the first expert system.

Second, you may not be the best judge of which problems are truly important. Cook was surprised by the widespread applicability of NP-completeness. Rabin and Scott left out some of their results on nondeterminism because they didn't want to make the paper too long. By his own description, Tarjan was just a graduate student looking for a problem when he started working on planarity testing. Backus thought Fortran might be useful for a single IBM machine model.

Third, you have to be the right person for the problem. As Donald Knuth puts it, "It's not true that necessity is the only mother or father of invention. . . . [A] person has to have the right background for the problem. I don't just go around working on every problem that I see. The ones I solve I say, oh, man, I have a unique background that might let me solve it—it's my destiny, my responsibility."

Fourth, great results demand a willingness to take risks and sometimes brave ridicule. McCarthy's first serious paper in artificial intelligence was denounced by an eminent mathematician as "half-baked." Most of Rabin's colleagues thought his use of randomization was a lucky trick that would work on few problems. Backus's proposal to make programming easier had to contend with the opposition of renowned physicist, mathematician, and computer designer John von Neumann, who believed that machine language programming was easy enough. Backus's 1978 proposal to replace Fortran-style programming languages with "functional" ones may never be accepted. Ditto for Lenat's Cyc project; Lenat notes, "You figure that as a researcher you will have on the order of three decade sized bets to make in your life."

Finally, what do computer scientists make of their discipline? How is it different from other sciences? Alan Kay puts it this way: "In natural science, Nature has given us a world and we're just to discover its laws. In computers, we can stuff laws into it and create a world."

Postscript

THE NEXT 25 YEARS

Our field is still in its embryonic stage. It's great that we haven't been around for 2000 years. We are still at a stage where very, very important results occur in front of our eyes.

—MICHAEL O. RABIN

hether you are a computer scientist or a computer user, the individuals profiled in this book have changed your lives. Their accomplishments set the pace for the principal advances in the field.

Fifty years ago, getting the hardware to work at all challenged the best engineers and physicists of the time.

Forty years ago, when several of our scientists began their careers, the challenge was to make programming easier. Researchers like Backus and McCarthy designed languages which made sense to scientists and to business and artificial intelligence programmers.

Thirty years ago, operating systems designers like Brooks and Dijkstra showed how a single computer could do many tasks simultaneously without forcing the programmer to know how the tasks are interacting. Thanks in part to Knuth, compilers simplified language design. Feigenbaum's expert systems proved that artificial intelligence could be practical.

Twenty years ago, Knuth, Rabin, Tarjan, Cook, and Levin showed how to do exact analysis of algorithms. Kay got programmers to think about using "objects" in software design, possibly eliminating the need to invent special-purpose languages for each new application. Lamport provided a framework for thinking about distributed systems.

Ten years ago, worldwide interest in artificial intelligence exploded and Lenat began working on Cyc. The wide availability of microprocessors encouraged Smith and Hillis to undertake two very different approaches to massively parallel design. Today people like Brooks apply parallel machines to solve problems from simulating molecules to creating virtual walkthrough kitchens.

Looking back, we can see a steady progression from a focus on the machine to a focus on problems. Processor design is now a minor challenge compared with designing the networks that efficiently link processors together. Software design that will make parallel processors behave like a single very fast and very reliable computer presents an even greater challenge. Although AI has not reached its grand goal, its partial successes have spun off new industries: new algorithms for graphics, visualization, and virtual reality help to make computers intelligence amplifiers. As they become indispensable, computers have become invisible. Automobile companies now put computer networks in their cars. Washing machines will soon talk to personal computers.

In the next 25 years, when computation is everywhere, what role will computer scientists play? Here we make some risky predictions.

- Algorithms are recipes for computer innovations, so they will remain the field's daily bread.

The P vs. NP question introduced by Cook and Levin will be resolved. If P = NP, public key cryptography will suffer a great blow, but many useful combinatorial problems will yield to algorithmic solutions. If P vs. NP is not resolved, the assumption that NP is larger than P will give rise to new heuristic algorithms, perhaps based on Hillis's evolutionary computing style.

No matter how the P vs. NP question is resolved, new cryptographic protocols will emerge if factorization yields to a randomized attack.

Interdisciplinary algorithms will be created for myriad new applications in such fields as optical telecommunications, gene design, virtual worlds, and robotic surgery.

- Like faces in photographs, new programming languages will always catch people's attention. But like beautiful images, the innovative and influential ones will remain rare.

Object-oriented programming languages such as Smalltalk, C++, and Eiffel will prevail over purely functional ones because the object-oriented paradigm models the natural evolution of products: if you build your product out of components, then the next version of the product can use some existing components and replace selective components. On the other hand, functional languages descendant from Backus's FP will find uses in applications requiring high reliability.

Specialized languages that can meld components written in different original languages will become popular.

- Computer science and robotics will reduce the time from design to manufacturing from months to days. The translation from electronic designs to steel and plastic will be nearly as easy as sending a fax.

  By the middle of the nineteenth century, the principles of Newtonian mechanics were well understood, and engineers and entrepreneurs brought forth innovations every year. We find ourselves now at the same stage in computer technology.

All computing will be connected, permitting wide-reaching information searches, but also rampant intellectual piracy. Ensuring privacy and data security while stimulating free communication will require fresh computer design (and legal) insights.

Reliability will become a dominant issue as software is used to mask hardware failures. This will make the recent work of Lamport, Dijkstra, Levin, and Cook on formal verification extremely relevant.

Just as computers have largely replaced wind tunnels in airplane wing design, so will they replace much experimentation in the physical and perhaps social sciences. This presents the danger that the experiments may no longer reflect nature, but the assumptions of the programmer.

Self-reproducing robotic factories for space will be designed and perhaps tested on earth.

Virtual reality will help designers test out their ideas, but will find its primary markets in entertainment and medicine.

Computing with gene sequences will find applications, especially in the food industry.

- Artificial intelligence will continue to spin off useful applications and problems, because AI researchers take the biggest risks.

Database systems will incorporate many of the ideas as well as the knowledge base from Lenat's Cyc or the descendants of that system.

Artificial intelligence will blend with virtual reality and neurophysiology in systems that will allow a person's brain waves to directly control his or her environment.

Education will reflect the growing computational IQ. Teachers won't bother to teach skills that computers can do better than humans, such as calculation and memorizing facts. In the best schools, students will solve problems in teams that draw upon information stores and expert systems available on networks. Reading, logical reasoning, and team skills will form the basis for a liberal education. Those few people who can learn uncodified specialized knowledge quickly will learn a field, enter their expertise into an expert system for a fee, and then move on.

We feel drawn to a certain image of the people featured in a book like this twenty-five years from now. Their flesh will be made of carbon, but their minds will be connected to computational limbs. They will use the computer to amplify their intelligence, while exercising their own very human creative genius. Godspeed.

Glossary

Note: Italicized words have a separate glossary entry.

access   either a *read* or a *write* to a computer's memory.

algorithm   a set of instructions that explains how to solve a problem. For example, a sorting algorithm must be able to take a large collection of names and place them in alphabetical order. An algorithm is usually described in English, using mathematical notation, precisely enough to be translated by an experienced programmer into a computer *program*.

amortization   a method of efficiency analysis invented by Robert Tarjan and Danny Sleator to measure the "worst case average" cost of a computation. Suppose you want to determine how many in an audience of 104 people have a last name beginning with A, beginning with B, and so on up to Z. You ask people to raise their hands when their letter is called. Suppose you measure cost by the number of hands you must count. Some letters like S will have a high cost, but other letters like Q will have a low cost. However, the amortized cost per letter is $^{104}/_{26} = 4$; the total cost is 104 hand countings divided by 26 letters in the alphabet.

ARPA   Advanced Research Projects Agency. A source of research grants sponsored by the Department of Defense. Most large projects described in this book received their funds from ARPA or (as it was called from the 1970s to the early 1990s) DARPA. ARPA also sponsored the research and development of the Arpanet which has grown into the Internet.

artificial intelligence (AI)   a subdiscipline of computer science with roots in philosophy, linguistics, and psychology. Its goal is to have a computer emulate intelligent human activities such as exercising expert knowledge, learning from its experience, or reasoning like a chess player.

artificial life   an interdisciplinary study involving physicists, biologists, and computer scientists whose goal is to use computer simulation to explore biological processes such as evolution, competition, and the nature of living complexity. Some work suggests that AL may also lead to a new paradigm for computing: state the problem, create artificial living cells, reward the cells for better solutions to the problem, and let the cells try to find a good answer.

assembly language   a symbolic representation of *machine language*. For example, to copy the contents of machine register AL to machine register CH on some Intel processors, one would type mov AL, CH in assembly language. The machine language representation is a much more forbidding 10001010 11000101.

atomicity   an operation O is atomic (literally indivisible) if any other operation can read the values of locations X and Y only either as they existed before O started or after O finished. The need for atomicity comes up in everyday life. For example, if Bob and Alice simultaneously put money in their joint bank account, the following sequence of events would lead to the loss of one of their deposits: Bob's program reads balance, Alice's program reads balance, Alice's program adds deposit value to balance, Bob's program adds his deposit value to the balance that his program read. This will cause all traces of Alice's deposit to be lost. However, if each deposit operation can be made atomic, then it will appear as if Bob's program executed before Alice's began or that Alice's program executed before Bob's began. Atomicity is usually achieved through *mutual exclusion*.

atomic register   any two operations on the *register* must appear to occur one at a time.

automatic programming   techniques to reduce programming to the minimum possible effort: just state the problem and let the computer do the rest. Popular in the 1960s, the field has periodically contributed to new ideas in programming languages, particularly modern-day specification languages.

axiom   a statement in a mathematical theory that is given without proof. For example, an axiom of Euclidean geometry states that the shortest path between two points is a line.

batch processing   having a computer execute one program from beginning to end without interference from any other program. By contrast, in multiprogramming or multitasking a computer does a little bit of one program, then a little bit of another, and so on. Batching was the first method used in computers, but then was largely supplanted by *multiprogramming*.

bit   a number that can take two values: 0 or 1. Eight bits make a *byte*.

Boolean algebra   a form of algebra invented by English logician George Boole in the 1850s to permit reasoning about propositional logic through calculation. True is represented by 1, false by 0, logical AND by multiplication, and logical OR by addition. (In Boolean algebra $1 + 1 = 1$.) If we know that X is true AND Y is false, then X AND Y is false, since true AND false is false. We would show this in Boolean algebra by noting that $1 \times 0 = 0$. Nearly a century later, Claude Shannon, most famous for the invention of information theory, would show that Boolean algebra was a good tool to analyze digital switching circuits.

breadth-first search   a technique for searching graph problems that is analogous to the following everyday algorithm for finding something in a book: look at all the chapter headings in the table of contents; if you don't find what you want, then look at the section headings; if you still don't find what you want, then read the text. See, for contrast, *depth-first search*.

buffer   part of a processor memory that is used for communication among processes running at different speeds. In everyday life, mailboxes act as buffers. We drop letters in the box at a variety of times and the letter carriers pick them up. If there were no mailbox-buffer, then we would have to wait for the letter carrier and give her the letters when we saw her. In the jargon of computer science, the person who puts data into the buffer is called the producer and the person who removes data is called the consumer.

byte   eight 1s and 0s in a sequence. A byte corresponds roughly to a single letter in a text file. A megabyte is a million bytes.

cache   small, fast memory (typically 100,000 bytes with access times of about 20 billionths of a second) used to speed up computer performance. Recently *accessed* data is copied from the larger main memory to the cache, where it replaces less recently accessed data. Similar to human short-term memory.

cache coherency   in a multiprocessor computer, many caches may have copies of the same data. If one processor changes that data, all the other caches must instantly be brought up to date to maintain equality, i.e., coherency.

combinatorics   the branch of mathematics concerned with finite sets and their relationships. Finding the quickest route from Cleveland to Chicago given travel times over specific roads is a combinatorial problem. Another one is finding a winning strategy in checkers given a certain position.

compiler/interpreter   the translator from a high-level language such as Fortran, C++, SQL, Excel, PhotoShop, or WordPerfect to machine language (the language of a processor). Compilers and interpreters allow people to write programs in an intelligible language without requiring the underlying processor to understand this language at all. Compilers produce faster code than interpreters because they perform the translation when the program is first written. However, interpreters are more flexible because they perform the translation at the moment of *execution*.

concurrency   a situation in which different activities overlap in time. Reading a novel at home may be concurrent with writing a report at work. That is, the time interval between the start and end of an activity overlaps with the time interval between the start and end of another activity.

conjunction   a set of assertions separated by logical AND. For example: Mary had a raincoat AND Mary had an umbrella.

consumer   see *buffer*.

context-free grammar   a type of grammar first described by linguist Noam Chomsky that is used for describing programming languages. The grammar consists of rules such as "a sentence is a noun phrase followed by a verb phrase" rendered as S → NP VP. Another rule might be "a noun phrase consists of an article followed by zero or more adjectives followed by a noun."

critical section   the part of a program that accesses data shared by many processes. *Mutual exclusion* algorithms must ensure that only one *process* executes its critical section at a time.

cryptography   literally, hidden or secret writing; the field concerned with sending messages intelligible to only the sender and receiver. Typically, the sender encrypts the message with a code and the receiver decrypts the message using a related code. In private-key cryptography, the receiver must know the sender's code. This may permit the receiver to masquerade as the sender. In public-key cryptography, the receiver does not know the sender's code and thus cannot pretend to be the sender. As currently practiced, public-key cryptography depends on the fact that it is much easier to multiply two large prime numbers than to find the prime factors of a very large number. If any security agency has figured out how to find prime factors of large numbers, we'd be the last to know.

DARPA Defense Advanced Research Projects Agency. See *ARPA*.

data base or database a large collection of data to support some human activity. For example, banks hold databases of account records; video stores hold databases of CDs and cassettes. Until the late eighties, large databases were stored in mainframes but parallel microcomputer-based databases seem to hold the most promise for the future.

data cleaning correcting errors in large *databases*.

dataflow architecture a computer design supporting a locally parallel form of computation. Most programs perform operations in a strict order determined by the program's text. In a dataflow program, an operation can take place as soon as the data it requires has been computed. A regular program resembles the instructions we read when assembling a model airplane kit: step 1, locate the pieces; step 2, glue together the fuselage, and so on. The instructions impose a strict but sometimes unnecessary order. A dataflow program resembles the instructions running through the head of an expert chef who is cooking several courses simultaneously: when the water boils, add the vegetables; when the cream is whipped, layer onto cake, etc.

data structure information storage method designed to facilitate rapid access. A dictionary is similar to a data structure: its alphabetical order facilitates rapid access to words. A persistent data structure allows quick retrieval of past data as well as current data. Imagine the usefulness of having access to every name and address you knew about five years ago, as well as those you know today.

deduction a form of internally consistent reasoning that is always valid. Philosopher Charles Sanders Peirce gave the following example of deduction: If I know that every bean in this bag is white and someone pulls a bean from the bag, then I know it's white. Contrast this with *induction*.

depth-first search a technique for searching graph problems that is analogous to the following everyday algorithm for finding something in a book: look at the first chapter, first section, and read it; then the first chapter, second section; and so on in order until you find what you want. For contrast see *breadth-first search*.

determinism precisely predicting the next state of a computation from its past states and the data it now accesses. By contrast, a nondeterministic computation may enter any one of several states in the same situation. "Solving" a problem nondeterministically in time T means that after T steps, one of the possible states is the correct one. Michael Rabin and Dana Scott showed that any nondeterministic program could be translated into a deterministic one, but nondeterministic programs are far easier to write.

dining philosophers problem   a problem invented by Edsger Dijkstra and named by Anthony Hoare that illustrates many of the principles and challenges of access to shared resources in a computer system. A group of philosophers are sitting around a table. Each has a plate of spaghetti in front of him. Between each pair of neighboring plates lies a chopstick. Thus the left chopstick of a philosopher is the right chopstick of his neighbor to the left. Each philosopher thinks for a while, eats for a while, goes back to thinking, and so on forever. To eat, a philosopher requires both chopsticks. The question is: Which program should the philosophers follow so that every philosopher eventually can eat? This simple problem and its hundreds of variants has been used as an explanatory vehicle for *mutual exclusion* techniques in multiprogrammed systems.

disambiguate   resolve the problem of multiple interpretations of a single sentence. For example, if you see the headline "Stolen Painting Found by Tree," you are meant to imagine that the painting was next to the tree when it was found. But if you read "Stolen Painting Found by Detective," then you might imagine a more active role for the detective. Disambiguating sentences requires knowledge of the world and common sense.

disjunction   a set of assertions separated by logical OR. For example: Mary had a raincoat OR Mary had an umbrella.

distributed algorithm   an *algorithm* that runs on a set of processors connected together by a network.

DO statements   in Fortran, these are statements that perform *loops*.

Euler tour   in a *graph*, an Euler tour is a route starting from a node that goes across every edge then returns to the starting node.

execution   the running of a program. If a recipe is analogous to a program, then cooking from a recipe is analogous to executing a program.

expert system   a computer program that simulates human expertise. Expertise here is defined broadly. For instance, an expert system may give medical advice about certain diseases or suggest how to remove a stuck drill bit from an offshore drilling hole.

exponential time   an unreasonably long time. To be specific, if an *algorithm* for a problem of size N takes time proportional to 2^N, then the algorithm is said to be exponential time. If the best possible algorithm for a problem is exponential time, then increasing the size of the problem by just one element doubles the time or often worse. An example of a exponential problem is listing all the possible combinations of letters of length N. When N is 1, then there are 26 possible combinations, from the 26 letters in the alphabet. When N is 2, there are 26×26 and so on. Even for the fastest conceivable computer, this problem becomes intractable once N is much above 10. Contrast this with *polynomial-time* algorithms.

factoring   finding the prime factors of a number (primes are numbers divisible only by themselves and 1). For example, the prime factors of 221 are 13 and 17. As far we know publicly (some security agencies might know more), factoring requires exponential time which is related to the number of digits in the number being factored (three decimal or 8 binary in the case of 221). If factoring were easy, then the best public-key *cryptography* scheme known would be easy to crack.

failure tolerance   also known as fault tolerance; the ability of a mechanism such as a network of computers to continue operating correctly even if some of the components fail.

Fermat's last theorem   a conjecture written in the margin of a manuscript: The equation $a^n + b^n = c^n$ has no solution for $n \geq 3$.

finite state automaton (plural automata)   a mathematical model of a machine that has a finite collection of "states" and transitions between states. Each transition is labeled with a letter. The states are arbitrary except there must be a starting state and one or more accepting states. If a sequence of letters is fed to a finite state automaton and the automaton ends in an accepting state when the input is exhausted, then the automaton is said to accept the input. A combination lock is a good example of a finite state automaton. Instead of following sequences of letters, dials are turned or buttons pushed. The accepting state corresponds to "lock opened." In a deterministic finite state automaton, there is one transition from every state for every letter. By contrast, a nondeterministic finite state automaton permits several transitions from a single state to share the same letter. Nondeterministic finite state automata are easy to program and can be translated to deterministic finite state automata. See *determinism*.

fixed point   numbers in a computer have a fixed length, nowadays usually equivalent to sixteen decimal digits. The fixed point computers of the early fifties represented numbers using these sixteen digits. The programmer had to remember that a 25 in this location meant 2.5 and that a 365 there meant 365,000,000,000,000,000. Few people other than John von Neumann found this intuitive. See *floating point*.

floating point   in answer to the problems of *fixed point* computation, hardware designers started to store exponents in computers. Thus, one could represent 2.5 as 25 E -1 and 2,500,000 as 25 E 5. John Backus's first major contribution to computer science was to design software that simulated floating point operations on a fixed point machine.

Fortran   computer language developed in the mid-1950s by a team at IBM led by John Backus. It was the first widely used computer language and remains the computer language of choice for many scientific applications.

functional programming   a style of programming in which results are computed from functions without any explicit notion of updating data in memory. Spreadsheets are like functional programs in that each cell has either a number or a function referring to other cells. John Backus designed a functional programming language he called FP.

grammar   a description of the syntax or acceptable form of a language. In English, the syntax of the following sentence, suggested by linguist Noam Chomsky, "Colorless green ideas sleep furiously," is correct, even though its meaning is obscure. Computer languages are normally described in a syntax called Backus-Naur form, which is essentially the same as Chomsky's notion of *context-free grammar*. Such grammars are too weak to describe human languages.

graph   in combinatorics and computer science, a collection of nodes (also called vertexes or vertices) and edges (also called arcs or links) that connect the nodes. A graph can represent an infinite variety of networks including a network of cities connected by roads or a network of pipes connecting oil drilling rigs to storage areas.

halting problem   the problem of determining whether machine X can decide whether another machine Y will ever stop if Y is asked to run program P on data D. Alan Turing, after describing a general computational device since called a Turing machine, showed that even his device could not guarantee a solution to the halting problem in all cases. That is, there must be instances of Y, P, and D for which X will not be able to determine a correct answer. In fact, this is only one of infinitely many "undecidable" problems only a few hundred of which have been found so far.

hardware the processors, disks, printers, monitors, and so on that make up a computer system. If you drop it on your foot, it will hurt.

heuristic a term coined by mathematician George Polya meaning a problem-solving technique that is not guaranteed to give the correct solution in all cases, but usually does. An everyday example: If the traffic light says walk, then do so without looking. This rule works well in London, but not as well in New York City.

induction reasoning from the particular to the general. To borrow from philosopher Charles Sanders Peirce's beanbag example: induction is the conjecture that because every bean you have taken from a bag so far is white, all beans in the bag are white. See *deduction*.

intelligence amplification augmenting human capabilities with the help of a computer. A specialized expert system does this when it helps a general practitioner apply the latest medical research to a patient with a rare disease. A virtual reality system does this when it allows an architect to walk through a building she has just designed.

interpreter/compiler See *compiler/interpreter*.

interrupt a signal to a processor from a device external to the processor. For example, every key typed on the keyboard of a personal computer "interrupts" the processor inside the computer. Interrupts are analogous to ringing telephones. Masking interrupts (one of Fred Brooks's innovations) permits the processor to shut off these interruptions for a time, thus avoiding electronic confusion.

Lisp a language for symbolic computation developed by John McCarthy in the 1950s. Lisp remains the main programming language of *artificial intelligence* and has influenced virtually every programming language developed since its invention.

loop a construct in a program that permits a certain set of instructions to be repeated many times. Most jobs can be described by loops. For example, the basic loop of a toll collector would be something like:

```
repeat until your next break
      greet driver
      collect money
      return change
      give directions if requested
end repeat
```

machine language the circuitry of every processor permits it to understand a certain set of instructions. These instructions are encoded into so-called binary machine language in the form of ones and zeros. For example, to copy data from register AL to CH on some Intel chips, one would type 10001010 11000101 in machine language. A compiler translates from a high-level language such as Fortran, Lisp, or Smalltalk to machine language. See *compiler/interpreter*.

mainframe multimillion-dollar computer for large organizations. Until 1990, IBM dominated this market, but collections of microprocessors interconnected by a network have taken over since then.

memory place where data is stored in a computer. There is a hierarchy among the memories: register memory is very fast but very small (access times of a few nanoseconds but storing 1000 bytes or less); cache memory is nearly as fast but much larger (hundreds of thousands of bytes); main memory, also known as random access memory (RAM), is larger but slower (millions of bytes in a hundred nanoseconds); disk memory is larger but slower still (billions of bytes in 15 milliseconds); tape or optical disk memory is the largest but the slowest (trillions of bytes requiring seconds). These numbers may change over time, but the basic relationships between size and speed will remain the same.

memory latency the time for a request to main memory to return with data. In current technology, a processor might be able to do 100 steps while it is waiting.

microprocessor a processor on a single chip having an area of about 4 square inches or 26 square centimeters. Current microprocessors cost anywhere from a few dollars to a few hundred dollars and are about as fast as multimillion-dollar machines of ten years ago.

microsecond millionth of a second.

millisecond thousandth of a second.

multiprocessor a computer having many processors connected to a common shared memory over a fast network.

multiprogramming having one processor execute several programs at once, each one being executed for a few milliseconds at a time. One benefit of this idea is that no single program can delay any other for too long. A second benefit is that while one program waits for data from a disk or the screen, another can do useful work. Contrast to *batch processing*.

multitasking   same as multiprogramming.

mutual exclusion   the assurance that two events cannot happen *concurrently*. Traffic lights attempt to ensure that cars moving in perpendicular directions don't go through the same point of an intersection at the same time.

nanosecond   billionth of a second.

nondeterminism   not *deterministic*.

nondeterministic polynomial (NP) problem   a problem for which any candidate solution can be tested quickly. (in the jargon: in polynomial time). Clearly, any problem that can be solved quickly (so-called P problems) also has this property. So, P problems are NP problems. The major open question in theoretical computer science is whether NP problems are also P problems.

NP-complete problem   a problem that is provably as difficult as any problem in NP. This means that if any NP-complete problem could be solved fast, then all problems in NP can be solved fast. A typical NP-complete problem is the traveling salesman problem: given a collection of cities and the costs of going from one city to the other, is there a route from, say, St. Louis that visits every city in the collection and returns to St. Louis costing less than $2000? A candidate solution for an NP-complete problem is easy to verify or reject quickly, but nobody knows whether a good solution can be found quickly in all cases. This means that NP-complete problems probably require exponential time.

object orientation   a style of programming in which each software component is treated as a black box; its external operations are visible to users of the components, but its internal implementation is hidden from view. In everyday life, radios are objects whose operations allow us to control volume and select frequencies, but we are warned not to look at the internal workings for fear of electrocution. Object orientation permits construction of extremely large programs from hopefully reliable and reusable components whose internal workings we never need to consider.

parallel computation   completing a task faster by having many processors work on parts of it. For example, a computer graphics program may assign each portion of a display screen to a different processor.

parallelizable task   a task that can be decomposed into subtasks that are more or less independent of one another. If a task is parallelizable, then it is faster to execute it on many processors than on a single one. In everyday life, the task consisting of boiling water for tea and baking a cake is parallelizable. Having a baby is not.

parsing a grammatical analysis. For natural languages like English, parsing a sentence corresponds to diagramming its parts of speech—articles, adjectives, nouns, verbs and so on—to determine whether the sentence is grammatically correct. For computer languages, parsing consists of determining whether a program is grammatically correct and how to interpret it. For example, in a programming language like Fortran, whose syntax is like that of normal arithmetic, x + y is grammatically correct, whereas x + ÷ y is not.

perebor a Russian word meaning "brute force." A perebor search is one that requires a search of all or nearly all possibilities to arrive at a solution. NP-complete problems are widely believed to require perebor.

persistent data structure see *data structure*.

pipeline parallelism a form of parallelism in which a processor simultaneously executes different operations at different stages of completion. In everyday life, a car assembly line implements pipeline parallelism. The line assembles many cars at once, but all are at different stages of construction. Consequently, the production rate of a pipeline or assembly line is limited by the time required to execute its longest step.

planarity testing determining whether a *graph* can be redrawn so that its edges meet only at nodes.

polynomial (P) problem a problem for which a fast *deterministic* solution exists. "Fast" for computer scientists means that the time to solve it is a polynomial function of its size as opposed to being an *exponential* function of its size. Specifically, if a problem of size N can be solved in time proportional to N^K for some fixed K independent of N, then it is polynomial (or "is in P"). Sorting, standard flow problems, and encryption are examples of polynomial-time algorithms.

prime number a number like 13, 17, and 101 that can be divided only by itself and 1. Prime numbers are used extensively in *cryptography*.

private-key cryptography see *cryptography*.

predicate calculus a formalism in logic for representing facts about groups as well as individuals, developed by the German mathematician Gottlob Frege in 1879 when he was 31. Predicate calculus permits statements such as: "All Olympic athletes are fit." If you believe that and are given the statement that Tonya is an Olympic athlete, you can infer that Tonya is fit. Thus, predicate calculus is more powerful than *propositional calculus,* which concerns only individuals.

process a running program. In everyday terms, if a processor is like a cook and a program is like a recipe, then a process is like a cook following a specific recipe.

processor a single computational element that can do operations like add, subtract, multiply, divide, and elementary logical operations. A computer may contain many processors.

producer see *buffer.*

program instructions given to a computer in a specific programming language so it performs some task. These are analogous to a recipe given to a cook in a human language and specified in full detail.

program verification also known as formal verification; the subdiscipline of computer science that attempts to prove mathematically that programs will do what they are supposed to do in all cases. A few researchers try to verify safety-critical software used for example in airplanes. The general industry practice is to run a collection of tests. If no bugs are found, the software is shipped.

programming language a means of instructing a computer what to do. There are hundreds of programming languages. Some, like *Fortran* and *Lisp,* are used primarily by scientists and engineers, but others like word processing languages are used by the rest of us. A good programming language is tailored to its application, just as a good shoe is tailored to an application (hiking, running, dancing, or beach). A *compiler* translates a high-level programming language to *machine language.*

propositional calculus a formalism in logic for representing facts about individuals. Rules for propositional calculus allow you to infer other propositions. For example, if the propositional sentences "Either Tweety is a bird OR Luke is a gazelle" AND "Luke is not a gazelle" are both true, then we can infer that "Tweety is a bird" is true. See *predicate calculus* for contrast.

public-key cryptography see *cryptography.*

random access memory (RAM) also known as main memory; memory having access times on the order of 100 billionths of a second. Most running programs are faster if they can stay in main memory rather than on disk.

randomized algorithm   a technique of algorithmic design pioneered by Michael Rabin in which a random process, such as an electronic coin flip, is used to find an answer. Such algorithms may sometimes arrive at an incorrect result (e.g., identifying a number to be prime that isn't), but do so with a probability that the algorithm designer can set. Typical probabilities are 1 in a trillion trillion.

read   a computer instruction that fetches the value of a location from computer memory into the processor without disturbing the value in the memory.

read-only memory (ROM)   a chip or a small collection of chips containing instructions that are inserted or "burned in" at the factory and then become unchangeable. The ROM typically contains instructions that bring in basic programs when a computer is turned on. In a human being, the nerves governing reflexes play a role analogous to a ROM.

recursion   a form of programming in which a function F is defined in terms of itself. In genealogy, recursion would be defining the term "ancestor" as either a parent or as the parent of an ancestor.

reducibility   a technique for showing that problem X is difficult to solve by showing that any solution for X would lead to a fast solution for Y (in the jargon, Y is reducible to X). If Y is known to be very hard, therefore X must be hard also.

registers   an element of very fast memory in a processor. Typically there are from 16 to 10,000 registers in a machine.

resolution rule   an efficient means for proving theorems in logic invented by John Alan Robinson.

semantics   a description of the meaning of a natural or computer language utterance. In natural language, a phrase may have meaningful semantics even though its *syntax* is incorrect (e.g., "Me go eat") Computer language utterances (i.e., programs) will not translate to *machine language* if their syntax is incorrect. If their syntax is correct but semantics are wrong (e.g., a statement adds instead of subtracts), the program may fail when used.

semaphore   a signaling mechanism that prevents trains from colliding. Edsger Dijkstra introduced this idea into computer science as a mechanism for ensuring *mutual exclusion*.

shortest-path problem   given a *graph* where each edge has a cost, the shortest-path problem finds the cheapest route from the source to the destination.

simulate behave in an analogous way. For example, a computer program might be written to simulate the flow of oil through sand. Such a program would permit geologists to pick the best place to drill for oil.

software one or more programs.

structured proofs a proof laid out hierarchically: the main assertions are at the first level, justifications for the first-level assertions are at the second level, justifications for the second-level assertions are at the third level, and so on.

syntax see *grammar.*

traveling salesman problem given a network of cities connected by airplane, the traveling salesman problem is to find the cheapest route allowing the salesman to visit all cities and return home. This problem is *NP-complete.*

transistor an electronic switch made up of semiconductor material, usually silicon or germanium. There are close to a billion transistors on a modern computer chip. These devices can store information or perform *Boolean* operations.

Turing test a test proposed in 1950 by Alan Turing to decide whether a computer exhibits human level intelligence. It goes like this: suppose a person typing at a terminal exchanges messages with a hidden interlocutor who is occasionally a computer and occasionally a human being. If the person cannot tell which is which (or who is who), then the computer has exhibited intelligent behavior. This test is one of the unsolved challenges of artificial intelligence.

undecidable problem a problem for which no computer program can guarantee a solution. See *halting problem* for an example.

vector an ordered collection of numbers whose interpretation depends on the application. In cartography, we represent points on the surface of the earth by a two-element vector representing latitude and longitude. Mecca is at (22, 40), whereas (40, 22) is not far from Athens.

vector processing the mathematical manipulation of (usually large) vectors. Scientific work often represents a physical phenomenon by a large vector whose elements give measurements at various points in space at a particular time. Predicting the evolution of the phenomenon requires calculation on that vector. These calculations are usually quite regular, permitting *pipeline parallel* processing.

virtual reality   a field pioneered by Fred Brooks, Ivan Sutherland, and others in which a computer creates an illusionary world in which a user can touch and control objects that have no external existence. In Brooks's work, a user can travel across the bonds of a molecule and pull atoms from their places.

write   a computer instruction that changes the value of a location in a computer memory.

References

We asked each scientist to select a small set of his own favorite papers and books. Those choices are listed by author. Following the authors' selections is a set of general references pertaining to each of the four parts of the book.

John W. Backus

Backus, John W. 1981. "The History of Fortran I, II, and III." *History of Programming Languages*. New York: Academic Press.

Backus, John W. 1978. "Can Programming Be Liberated from the von Neumann Style? A Functional Styles and Its Algebra of Programs." *Communications of the ACM* 21(8):613–641.

Backus, John W., R. J. Beeber, S. Best, R. Goldberg, L. M. Haibt, H. L. Herrick, R. A. Nelson, D. Sayre, P. B. Sheridan, H. Stern, I. Ziller, R. A. Hughes, and R. Nutt. 1957. "The Fortran Automatic Coding System." *1957 Western Joint Computer Conference*, pp. 188–198.

Frederick P. Brooks, Jr.

Airey, John, John Rohlf, and Frederick P. Brooks, Jr. 1990. "Towards Image Realism with Interactive Updates in Complex Virtual Building Environments." Proceedings of 1990 Symposium on Interactive 3D Graphics, *Computer Graphics* 24(2):41–50.

Amdahl, G. M., G. A. Blaauw, and F. P. Brooks, Jr. 1964. "Architecture of the IBM System/360." *IBM Journal of Research and Development* 8:87–101.

Bergman, Lawrence D., Jane S. Richardson, David C. Richardson, and F. P. Brooks, Jr. 1993. "VIEW—An Exploratory Molecular Visualization

System with User-Definable Interaction Sequences." *Computer Graphics: Proceedings of SIGGRAPH 93,* 27(4):117–126.

Blauww, G. A., and Brooks, F. P., Jr. *Computer Architecture* (2 vols.). Reading, MA: Addison-Wesley, in preparation.

Brooks, F. P., Jr. 1988. "Grasping Reality through Illusion: Interactive Graphics Serving Science." In D. Frye and S. Sheppard (eds.), *Computer Human Interaction '88 Proceedings.* Reading, MA: Addison-Wesley, pp. 1–11.

Brooks, F. P., Jr. 1987. "No Silver Bullet—Essence and Accidents of Software Engineering." *IEEE Computer* 20(4):10–19.

Brooks, F. P., Jr. 1977. "The Computer 'Scientist' as Toolsmith: Studies in Interactive Computer Graphics." In B. Gilchrist (ed.) *Information Processing 77.* Amsterdam: North-Holland.

Brooks, F. P., Jr. 1975. *The Mythical Man-Month: Essays on Software Engineering.* Reading, MA: Addison-Wesley.

Brooks, Frederick P., Jr., Ming Ouh-Young, James J. Batter, and P. Jerome Kilpatrick. 1990. "Project GROPE: Haptic Displays for Scientific Visualization." *Computer Graphics: Proceedings of SIGGRAPH 90,* 24(4): 177–185.

Taylor, Russell M. II, Warren Robinett, Vernon L. Chi, Frederick P. Brooks, Jr., William V. Wright, R. Stanley Williams, and Erik J. Snyder. 1993. "The Nanomanipulator: A Virtual-Reality Interface for a Scanning Tunneling Microscope." *Computer Graphics: Proceedings of SIGGRAPH 93,* 21(4):127–134.

Stephen Cook

Aanderaa, S. O., and Stephen Cook. 1969. "On the Minimum Computation Time of Functions." *Transactions of the American Mathematical Society* 142:291–314.

Borodin, A., and Stephen Cook. 1982. "A Time-Space Tradeoff for Sorting on a General Sequential Model of Computation." *Siam Journal on Computing* 11(2):287–297.

Cook, Stephen. 1975. "Feasibly Constructive Proofs and the Propositional Calculus." *Proceedings of the Seventh Annual ACM Symposium on the Theory of Computing,* May 1975, pp. 83–97.

Cook, Stephen. 1971. "The Complexity of Theorem Proving Procedures." *Proceedings of the Third Annual ACM Symposium on the Theory of Computing,* May 1971; pp. 151–158.

Cook, S., and R. Reckhow. 1979. "The Relative Efficiency of Propositional Proof Systems." *Journal of Symbolic Logic* 44(1):36–50.

Edsger W. Dijkstra

Dijkstra, Edsger W. 1974. "Self-Stabilizing Systems in Spite of Distributed Control." *Communications of the ACM* 17:453–455.

Dijkstra, Edsger W. 1960. "Recursive Programming" *Numerische Mathematik* 2:312–318.

Dijkstra, Edsger W., and Carel A. Scholten. 1990. "Predicate Calculus and Program Semantics." In *Texts and Monographs in Computer Science.* New York: Springer-Verlag.

Dijkstra, Edsger W., and Carel A. Scholten. 1980. "Termination Detection for Diffusing Computations." *Information Processing Letters* 11(1):1–4.

Edward A. Feigenbaum

Feigenbaum, E. A. 1977. "The Art of Artificial Intelligence: Themes and Case Studies in Knowledge Engineering." *International Joint Conference on Artificial Intelligence* 1977.

Feigenbaum, E. A., and J. Feldman (eds.). 1963. *Computers and Thought.* New York: McGraw-Hill.

Feigenbaum, E. A., B. Buchanan, and J. Lederberg. 1971. "On Generality and Problem Solving: A Case Study Using the DENDRAL Program." In D. Michie (ed.), *Machine Intelligence 6.* New York: Elsevier, 165–190.

W. Daniel Hillis

Hillis, W. D. 1993. "Why Physicists Like Models and Why Biologists Should." *Current Biology* 3(2).

Hillis, W. D. 1988. "Intelligence As an Emergent Behavior: Or, The Songs of Eden." *Daedalus* 117(1):175–189.

Hillis, W. D. 1985. *The Connection Machine.* Cambridge: The MIT Press.

Hillis, W. D., and B. M. Boghosian. 1993. "Parallel Scientific Computation." *Science* 13:856–863.

Hillis, W. D., and Guy L. Steele. 1986. "Data Parallel Algorithms." *Communications of the ACM* 29(12):1170–1173.

Hillis, W. D., and Lewis W. Tucker. 1993. "The CM-5 Connection Machine: A Scalable Supercomputer." *Communications of the ACM* 36(11):30–40.

Alan C. Kay

Kay, Alan. 1993. "The Early History of Smalltalk," *ACM SIGPLAN,* March 1993, pp. 69–96.

Kay, Alan. 1991. "Computers, Networks and Education." *Scientific American,* September 1991, pp. 138–148.

Kay, Alan. 1984. "Computer Software." *Scientific American,* September 1984, pp. 52–59.

Kay, Alan. 1977. "Microelectronics and the Personal Computer." *Scientific American,* September 1977, pp. 230–239.

Kay, Alan. 1972. "A Dynamic Medium for Creative Thought." NCTE Conference, November 1972.

Kay, Alan. 1972. "A Personal Computer for Children of All Ages." ACM National Conference, August 1972.

Kay, Alan, and Adele Goldberg. 1977. "Dynamic Personal Media." *IEEE Computer,* March 1977, pp. 31–42.

Kay, Alan, and Adele Goldberg. 1976. Smalltalk-72 Instruction Manual. Xerox PARC, March 1976.

Donald E. Knuth

Graham, Ronald L., Donald E. Knuth, and Oren Patashnik. 1989. *Concrete Mathematics.* Reading, MA: Addison-Wesley.

Knuth, Donald E. 1991. *Literate Programming.* Stanford, CA: Center for the Study of Language and Information.

Knuth, Donald E. 1986. *Computers and Typesetting.* (series.) Reading, MA: Addison-Wesley, 1986.

Knuth, Donald E. 1968. "Semantics of Context-Free Languages." *Mathematical Systems Theory* 1968:127–145; 1971:95–96.

Knuth, Donald E. 1968. *The Art of Computer Programming.* (series.) Reading, MA: Addison-Wesley.

Knuth, Donald E. 1965. "On the Translation of Languages from Left to Right." *Information and Control* 1965:607–639.

Knuth, Donald E., and Peter B. Bendix. 1970. "Simple Word Problems in Universal Algebras." in J. Leech (ed.) *Computational Problems in Abstract Algebra.* New York: Pergamon.

Knuth, Donald E., Svante Janson, Tomasz Luczak, and Boris Pittel. 1993. "The Birth of the Giant Component." *Random Structures and Algorithms* 1993:233–358.

Leslie Lamport

Lamport, Leslie. 1987. "A Fast Mutual Exclusion Algorithm." *Transactions on Computer Systems* 5(71):1–11.

Lamport, Leslie. 1979. "How to Make a Multiprocessor Computer That Correctly Executes Multiprocess Programs." *IEEE Transactions on Computers* C-28(9):690–691.

Lamport, Leslie. 1978. "Time, Clocks, and the Ordering of Events in a Distributed System." *Communications of the ACM* 21(7):558–565.

Lamport, Leslie. 1977. "Concurrent Reading and Writing." *Communications of the ACM* 20(11):806–811.

Lamport, Leslie. 1974. "A New Solution of Dijkstra's Concurrent Programming Problem." *Communications of the ACM* 17(8):453–455.

Douglas B. Lenat

Lenat, Douglas. 1984. "Computer Software for Intelligent Systems: An UnderView of AI." *Scientific American,* September 1984, pp. 204–213..

Lenat, Douglas. 1983. "Three Case Studies in Learning." In R. S. Carbonell, J. G. Michalski, and T. M. Mitchell (eds.) *Machine Learning.* Palo Alto, CA: Tioga Press.

Lenat, Douglas. 1977. "The Ubiquity of Discovery." *The Journal of Artificial Intelligence,* December 1977, pp. 257–285.

Lenat, Douglas. 1975. "BEINGS: Knowledge as Interacting Experts." *Proceedings of the Fourth International Joint Conference on Artificial Intelligence 75.* Tbilisi, U.S.S.R, September 1975.

Lenat, D., D. Borning, D. McDonald, S. Taylor, and S. Weyer. 1983. "Knowsphere: Design of an Expert System with an Encyclopedic Knowledge Base." *Proceedings of the Eighth International Joint Conference on Artificial Intelligence 83.* Karlsruhe, Germany, August 1983.

Lenat, Douglas, and John Seely Brown. 1984. "The Nature of Heuristics IV: Why AM and Eurisko Appear to Work." *Journal of Artificial Intelligence,* July 1984.

Lenat, Douglas, and Edward Feigenbaum. 1991. "On the Thresholds of Knowledge." *Artificial Intelligence* 47:185–250.

Lenat, Douglas, and R. V. Guha. 1994. "Enabling Agents to Work Together." *Communications of ACM,* July 1994, pp. 126–142.

Lenat, Douglas, and R. V. Guha. 1990. *Building Large Knowledge-Based Systems: Representation and Inference in the CYC Project.* Reading, MA: Addison-Wesley.

Leonid Levin

Babai, L., L. Fortnow, L. Levin, and M. Szegedy. 1991. "Checking Computations in Polylogarithmic Time." *ACM Symposium on the Theory of Computing,* 1991, pp. 21–31.

Goldreich, Oded, and Leonid Levin. 1989. "A Hard-Core Predicate for All One-Way Functions." ACM *Symposium on the Theory of Computing,* 1989, pp. 25–32.

Levin, Leonid. 1964. "On Storage Capacity for Algorithms." *Doklady Akademii Nauk SSSR* 14(5):1464–1466.

Levin, Leonid. 1984. "Randomness Conservation Inequalities." *Information and Control* 61(1):15–37.

Venkatesan, R., and Leonid Levin. 1984. "Random Instances of a Graph Coloring Problem Are Hard." ACM *Symposium on the Theory of Computing,* 1984, pp. 217–222.

John McCarthy

McCarthy, John. 1993. "Notes on Formalizing Context." *International Joint Conference on Artificial Intelligence—93,* pp. 555–560.

McCarthy, John. 1986. "Applications of Circumscription to Formalizing Common Sense Knowledge." *Artificial Intelligence,* April 1986, pp. 89–116.

McCarthy, John. 1960. "Recursive Functions of Symbolic Expressions and their Computation by Machine." *Communications of the ACM,* April 1960, pp. 184–194.

McCarthy, John. 1959. "Programs with Common Sense." *Proceedings of the Teddington Conference on the Mechanization of Thought Processes.* London: Her Majesty's Stationery Office.

McCarthy, John, and P. J. Hayes. 1969. "Some Philosophical Problems from the Standpoint of Artificial Intelligence." In D. Michie (ed.) *Machine Intelligence 4.* New York: Elsevier.

Michael O. Rabin

Rabin, Michael. 1989. "Efficient Dispersal of Information for Security, Load Balancing, and Fault Tolerance." *Journal of ACM* 38:335–348.

Rabin, Michael. 1979. "Digital Signatures and Public-Key Functions Are as Intractable as Factorization." *MIT Laboratory for Computer Science Technical Report no. 212.*

Rabin, Michael. 1976. "Probabilistic Algorithms." In J. F. Traub (ed.). *Algorithms and Complexity: New Directions and Recent Trends.* New York: Academic Press, 1976.

Rabin, Michael. 1969. "Decidability of Second-Order Theories and Automata on Infinite Trees." *Transactions of the American Mathematical Society* 141:1–35.

Rabin, Michael. 1963. "Probabilistic Automata." *Information and Control* 6:230–245.

Rabin, Michael. 1960. "Degree of Difficulty of Computing a Function and a Partial Ordering of Recursive Sets." *Office of Naval Research Contracts Technical Report*. Jerusalem: Hebrew University.

Rabin, Michael. 1958. "Recursive Unsolvability of Group Theoretic Problems. *Annals of Mathematics* 67:172–194.

Rabin, Michael, Y. Aumann, Z. M. Kedem, and K. V. Palem. 1993. "Highly Efficient Asynchronous Execution of Large Grained Parallel Programs." *34th Symposium on the Foundations of Computer Science, 1993*, pp. 271–280.

Rabin, M., and M. Fischer. 1974. "Super Exponential Complexity of Presburger Arithmetic." *SIAM-AMS Proceedings of Symposium on Complexity of Computations* 7:27–41.

Rabin, M., and R. Karp. 1987. "Efficient Randomized Pattern-Matching Algorithms." *IBM Journal of Research and Development* 31:249–260.

Rabin, Michael, and Dana Scott. 1959. "Finite Automata and Their Decision Problems." *I.B.M. Journal of Research and Development* 3:114–125.

Burton J. Smith

Alverson, Robert, David Callahan, Daniel Cummings, Brian Koblenz, Allan Porterfield, and Burton Smith. 1990. "The Tera Computer System." *Proceedings 1990 International Conference on Supercomputing, Amsterdam, June 1990*, pp. 1–6.

Cathey, W. T., and Burton Smith. 1979. "High Concurrency Data Bus Using Arrays of Optical Emitters and Detectors." *Applied Optics* 18:1687.

Smith, Burton. 1990. "The End of Architecture." Keynote address presented at the 17th International Conference on Computer Architecture Seattle, Washington, December 1990. *Architecture News* 18(4):10.

Smith, Burton. 1987. "Shared Memory, Vectors, Message Passing, and Scalability." *Proceedings 1987 DFVLR Seminar on Parallel Computing in Science and Engineering, Lecture Notes in Computer Science 295*. New York: Springer-Verlag, pp. 29–34.

Smith, Burton. 1985. "The Architecture of HEP." In J.S. Kowalik (ed.). *Parallel MIMD Computation: The HEP System and Its Applications*. Cambridge, MA: MIT Press, pp. 41–58.

Smith, Burton, and David Callahan. 1990. "A Future-Based Parallel Language for a General-Purpose Highly Parallel Computer." In

D. Gelernter, A. Nicolau, and D. Padua (eds.). *Languages and Compilers for Parallel Computing,* Research Monographs in Parallel and Distributed Computing. Cambridge, MA: Pitman and MIT Press, pp. 95–113.

Robert E. Tarjan

Goldberg, Andrew V., and Robert E. Tarjan. 1988. "A New Approach to the Maximum-Flow Problem." *Journal of the ACM* 35:921–940.

Hopcroft, John, and Robert E. Tarjan. 1974. "Efficient Planarity Testing." *Journal of the ACM* 21(4):549–568.

Lipton, Richard J., and Robert E. Tarjan. 1979. "A Separator Theorem for Planar Graphs." *SIAM Journal of Applied Mathematics* 36:177–189.

Sarnak, Neil, and Robert E. Tarjan. 1986. "Planar Point Location Using Persistent Search Trees" *Communications of the ACM* 29:669–679

Tarjan, Robert E. 1987. "Algorithm Design." *Communications of the ACM* 30:205–212.

Tarjan, Robert E. 1985. "Amortized Computational Complexity." *SIAM Journal of Algorithms and Discrete Methods* 6:306–318.

Tarjan, Robert E. 1983. *Data Structures and Network Algorithms.* Philadelphia: Society for Industrial and Applied Mathematics.

Tarjan, Robert E. 1972. "Depth-First Search and Linear Graph Algorithms." *SIAM Journal on Computing* 1972:146–160.

References for Part 1: Linguists

Chomsky, Noam. 1957. *Syntactic Structures.* The Hague: Mouton.

Gelernter, David, and Suresh Jagannathan. 1990. *Programming Linguistics.* Cambridge, MA: MIT Press.

Goldstine, H. 1972. *The Computer from Pascal to von Neumann.* Princeton, NJ: Princeton University Press.

Sethi, Ravi. 1990. *Programming Languages: Concepts and Constructs.* Reading, MA: Addison-Wesley.

References for Part 2: Algorithmists

Aho, A. V., J. E. Hopcroft, and J. D. Ullman. 1971. *The Design and Analysis of Computer Algorithms.* Reading, MA: Addison-Wesley.

Cormen, T. H., C. E. Leiserson, and R. L. Rivest. 1991. *Introduction to Algorithms.* New York: McGraw-Hill.

Davis, Martin. 1987. "Mathematical Logic and the Origin of Modern Computers." *Studies in the History of Mathematics*. Washington, D.C.: The Mathematical Association of America.

Garey, Michael R., and David. S. Johnson. 1979. *Computers and Intractability: A Guide to the Theory of NP-Completeness*. New York: W. H. Freeman.

Herken, Rolf (ed). 1988. *The Universal Turing Machine: A Half Century Survey*. Oxford: Oxford University Press.

Hodges, A. 1983. *Alan Turing: The Enigma*. New York: Simon and Schuster.

Kolmogorov, A. N., and V. A. Uspenskii. "Algorithms and Randomness." *Theoria Veroyatnostey i ee Primeneniya* (Theory of Probability and Its Applications) 3(32):389–412.

Trahktenbrot, B. A. 1984. "A Survey of Russian Approaches to Perebor (Brute-Force Search) Algorithms." *Annals of the History of Computing* 6:384–400.

References for Part 3: Architects

Flynn, M. J. 1972, "Some Computer Organizations and their Effectiveness." *IEEE Transactions on Computers* C-21(9):948–960.

Friedman, D. P., and D. S. Wise. 1978. "Aspects of Applicative Programming for Parallel Processing." *IEEE Transactions on Computers* C-27(4):289–296.

Jefferson, D. R. 1985. "Virtual Time." *ACM Transactions on Programming Languages and Systems* 7(3):404–425.

Ladner, R. E., and M. J. Fischer. 1980. "Parallel Prefix Computation." *Journal of the ACM* 27(4):831–838.

Schwartz, J. 1. 1980. "Ultracomputers." *ACM Transactions on Programming Languages and Systems* 2(4):484–521.

References for Part 4: Sculptors of Intelligent Machines

Crevier, Daniel. 1993. *AI: The Tumultuous History of the Search for Artificial Intelligence*. New York: Basic Books.

Dreyfus, Hubert L. 1979. *What Computers Can't Do*. New York: Harper & Row.

Levy, Steve. 1992. *Artificial Life: A Report from the Frontier Where Computers Meet Biology*. New York: Vintage.

McCorduck, Pamela. 1979. *Machines Who Think*. San Francisco: W. H. Freeman.

Michakski, Ryszard (ed.). 1983. *Machine Learning—An AI Approach*. Palo Alto, CA: Tioga Publishing.

Minsky, Marvin. 1985. *The Society of Mind*. New York: Simon and Schuster.

Moravec, Hans. 1988. *Mind Children: The Future of Robot and Human Intelligences*. Cambridge, MA: Harvard University Press.

Schank, Roger C. 1987. *The Cognitive Computer: On Language, Learning, and Artificial Intelligence*. New York: Walker.

Yazdani, M., and A. Narayanan (eds.) 1986. *Artificial Intelligence—Human Effects*. Chichester, England: E. Horwood.

Acknowledgments

The computer scientists profiled in this book gave generously of their time and knowledge during long interviews and their careful review of our first drafts.

Colleagues and students, especially Brad Barber, Ernie Davis, Martin Davis, Ben Goldberg, Mordecai Golin, Andrea Koenig, Lisa M. List, Ed Schonberg, Jack Schwartz, and Matthew Smosna helped make the manuscript more coherent. Also, loyal friends such as Albert and Naoko Adams, Amanda Brauman, Hadass Harel, and Martha Palubniak offered many useful suggestions. Linda Houseman and Herbert Stoyan provided us with valuable photographs. Loren Singer remains a constant source of writer's wisdom.

Sylvia Warren was an outstanding copy editor—precise, perceptive, and constructive.

As in the past, Jerry Lyons has been a delightful publisher who quietly suggested new directions for us to explore. Our assistant editor Liesl Gibson was always helpful and supportive, as were our supervising production editor Steven Pisano and design supervisor Karen Phillips. We also thank the proofreader, Jacquie Edwards.

Index